Anthropocide

Through an examination of Alfonso Cuarón's *Children of Men*, this book demonstrates the ability of cinematic fictions, and other complex narrative fictions, to contribute to meeting the climate challenge by shaping the desires of audiences.

What if there was a single feature film that showed us everything we need to know about climate catastrophe culture? What if that same film also made the philosophies of Slavoj Žižek, Mark Fisher, Francis Fukuyama, and Fredric Jameson accessible? Identifying the climate challenge as a cultural challenge, this book provides an unprecedented criminological analysis of both *Children of Men* and Fisher's *oeuvre* from 1998 to 2022 and demonstrates the capacity of cinematic narratives to shape climate catastrophe culture. Seeking to be part of the solution to the climate challenge, it is the first criminological study to link the capacity of cinematic fictions to shape desire to solutions to the climate crisis. It is also one of the most detailed and most rigorous criminological case studies of a cinematic work to date.

Anthropocide: An Essay in Green Cultural Criminology will be of great interest to students and scholars of green criminology, cultural criminology, narrative criminology, film theory, philosophy of film, and ecocriticism.

Rafe McGregor is Reader in Criminology at Edge Hill University, Lancashire, England. A critical theorist with research expertise in culture, political violence, and policing, he is the author of 20 books, including *Anthropocide: An Essay in Green Cultural Criminology* (2025), *Recovering Police Legitimacy: A Radical Framework* (2024), and *Literary Theory and Criminology* (2023). He is a fellow of the Royal Society of Arts and of Advance HE.

Routledge Studies in Crime, Culture and Media

Routledge Studies in Crime, Culture and Media offers the very best in research that seeks to understand crime through the context of culture, cultural processes and media.

The series welcomes monographs and edited volumes from across the globe, and across a variety of disciplines. Books will offer fresh insights on a range of topics, including news reporting of crime; moral panics and trial by media; media and the police; crime in film; crime in fiction; crime in TV; crime and music; 'reality' crime shows; the impact of new media including mobile, Internet and digital technologies, and social networking sites; the ways media portrayals of crime influence government policy and lawmaking; the theoretical, conceptual and methodological underpinnings of cultural criminology.

Books in the series will be essential reading for those researching and studying criminology, media studies, cultural studies and sociology.

America's Horror Stories
U.S. History Through Dark Tourism
Kevin Revier and Favian Alejandro Martín

Sexual and Physical Violence in Australian Punk and Hardcore Music Scenes
Ash Barnes

Human Trafficking Hysteria
Historical and Modern Perspectives on Moral Panics, Media, and Crime
Sarah Hupp Williamson

Anthropocide
An Essay in Green Cultural Criminology
Rafe McGregor

Anthropocide

An Essay in Green Cultural Criminology

Rafe McGregor

LONDON AND NEW YORK

First published 2025
by Routledge
4 Park Square, Milton Park, Abingdon, Oxon OX14 4RN

and by Routledge
605 Third Avenue, New York, NY 10158

Routledge is an imprint of the Taylor & Francis Group, an informa business

British Library Cataloguing-in-Publication Data
A catalogue record for this book is available from the British Library

ISBN: 978-1-032-93424-2 (hbk)
ISBN: 978-1-032-93430-3 (pbk)
ISBN: 978-1-003-56582-6 (ebk)

DOI: 10.4324/9781003565826

Typeset in Times New Roman
by Apex CoVantage, LLC

Contents

Acknowledgements

I would like to thank Avi Brisman and Nigel South, the pioneers of green cultural criminology, for their genius, generosity, and open-mindedness.

I would also like to thank Stephen Theaker for allowing me to test several of the ideas in this essay in his marvellous magazine, *Theaker's Quarterly Fiction*, in the form of short essays on climate change cinema and cultures of climate change and long reviews of *Children of Men* and both seasons of *Carnival Row*.

Fredric Jameson (1934–2024) was one of the world's greatest critical theorists. He died on 22 September 2024, the week I finished the first draft of this essay. It is too late to thank him for his inspiration, but I leave it here as a matter of record.

This essay was written with the support and funding of the Croatian Science Foundation as part of the research project *Aesthetic Education through Narrative Art and its Relevance for the Humanities* (UIP-2020-02-1309).

Part I

The Problem

1 Crisis

My aim in this essay is to convince you, the reader, that cinematic fictions have the capacity to confront the climate challenge by means – or, in virtue – of their capacity to shape the desires of audiences. This is not, like my book on literary theory, an essay *about* desire but an essay *about* the ways in which desire is shaped by films (McGregor 2023). I hope to achieve my aim by breaking my core claim into three constitutive claims, which correspond to the three parts of the essay. First (Part I), that the climate challenge is a cultural rather than natural challenge (or, more accurately speaking, both). Second (Part III), that cinematic and other complex narrative fictions can meet that challenge. And third (Part II), that Alfonso Cuarón's Children of Men (2006) is exemplary with respect to meeting the challenge of climate change. This chapter begins Part I, which is titled The Problem, where the problem is not only the challenge of climate change but also climate change as a cultural challenge. As such, I start with the science of climate change, move on to the most compelling social scientific explanation of the situation described by climatologists, and conclude with an introduction to criminology's role in attempting to control, reduce, or prevent the harms associated with climate change.

1.1 Climate Change

The United Nations (UN) Intergovernmental Panel on Climate Change (IPCC 2025) was established by the World Meteorological Organization and the UN Environment Programme in 1988. The mandate of the IPCC is to research anthropogenic climate change for the purpose of providing governments with a scientific basis for their climate policies. Research is undertaken in rolling assessment cycles of five to seven years at the end of which a series of reports are published, the first of which was released in 1990. As of 2024, the IPCC (2024) is in its seventh cycle, which will conclude in 2029. There are four main reports published in each cycle – *The Physical Science Basis*; *Impacts, Adaptation and Vulnerability*; *Mitigation of Climate Change*; and *Synthesis* – the first three of which are several thousand pages in length. The

DOI: 10.4324/9781003565826-2

IPCC does not conduct its own primary research, and the findings in the reports are reached by means of systematic reviews and meta-analyses of both original research and previous assessment reports. The panel has a reputation for rigour and conservatism, meaning that the findings are very likely accurate and that if they are inaccurate, they very likely understate rather than overstate the present and future impacts of climate change. My source for this essay is the Sixth Assessment Report (IPCC 2021, 2022a, 2022b, 2023), the components of which were published from 2021 to 2023.

There is almost complete agreement that the Earth's average surface air temperature has increased by between 1.1°C and 1.2°C since 1900 and that more than half of that increase has occurred since the mid-1970s (The Royal Society 2025). There is complete agreement that human activity has contributed to the rise in air surface temperature, although the full extent of that contribution is debated (IPCC 2021). There is widespread agreement that the consequences of air surface temperature increase are a rise in ocean level and acidity and more numerous and severe floods, droughts, and hurricanes (Emanuel 2018). The Sixth Assessment Report predicts significant and irreversible environmental and social harm once global warming reaches 1.5°C above pre-industrial temperatures. The IPCC (2022a: 7) is clear that 'there is at least a greater than 50% likelihood that global warming will reach or exceed 1.5°C in the near-term,' where 'near-term' is defined as *by 2040*. This increase can only be avoided by significant reductions in greenhouse gas emissions, including the reduction of CO_2 emissions to net zero, that is by significant changes to human activity. In *Climate Change Criminology*, Rob White (2018: 3) describes the evidence for climate change as compelling, comprehensive, and consistent and states that 'most disagreement surrounding climate change today is over how quickly global warming is proceeding rather than over whether it is happening.' I take scepticism about the scientific consensus as instances of what Avi Brisman (2012: 43) calls 'contrarianism,' that is disagreement on the basis of ideological rather than evidential grounds (in much the same way as I take scepticism about biological evolution).

Bruce Glavovic, Timothy Smith, and Iain White (2022: 829) are unequivocal about the status of the evidence:

> Governments concur that the science is settled on the reality of global change. Consensus dates back at least to the 1972 Stockholm Conference, was reiterated at the 1992 and 2002 Earth Summits, and in subsequent global agreements, including the United Nations Framework Convention on Climate Change (UNFCCC) 2015 Paris Agreement and Aichi Biodiversity Targets (2011–2020).

They argue that it is irresponsible for the IPCC to conduct its Seventh Assessment because it has already achieved its aim by establishing a global consensus on the science of climate change. Given that governments demonstrated their capability in acting on scientific evidence during the COVID-19 pandemic but have failed to act on the scientific evidence for anthropogenic climate change, Glavovic, Smith, and White contend that scientists should focus their efforts on exposing and explaining this failure. As drastic a step as this would be to take, they (Glavovic, Smith & White 2022: 832) believe it is the 'only effective way to arrest the tragedy of climate change science.' What they refer to as the *tragedy* of climate change science is the basis of the claim I shall make throughout this essay, that the challenge posed by climate change is a cultural rather than a natural challenge. Of course, the challenge is natural *and* cultural, but the natural challenge has already been met with the IPCC's identification of both the problem and solutions to that problem. The continuing challenge is the implementation of those solutions, which is a cultural problem. I shall not say a great deal more about the natural challenge of climate change because, first, I am not a natural scientist and, second, I agree with Glavovic, Smith, and White that the way to meet the natural challenge is to take the IPCC's Sixth Assessment Report seriously, that is use the scientific evidence as the basis for meaningful action.

Climate change is both anthropogenic and ecocidal. The IPCC has provided – and continues to – provide evidence for the first of these characteristics. The late Polly Higgins (2010: 63) defined ecocide as 'extensive damage to, destruction of or loss of ecosystem(s) of a given territory.' Rob White (2018) and David Whyte (2020) augment the legal definition with a zemiological one, as *environmental harm on a global scale* and *the threat to planetary sustainability* respectively. White also describes ecocide as *anthropocentric ecocide*, *genocide through geocide*, and *the crime of the century* (all of which I have paraphrased). I define *ecocide* as the reduction of the planet's capacity to support human and other animal life. The IPCC's predictions are that climate change is ecocidal. I (McGregor 2023: 152) previously described ecocide thus defined as 'anthropocidal,' an adjective selected to denote the potential for ecocide to destroy the entire human species (as well as many nonhuman species). Ecocide is, of course, neither actually anthropocidal nor apocalyptic. Nonhuman animal life has already survived five mass extinctions and will survive the sixth, evolving as it adapts to the warmer climate in a reversal of what happened in the Ice Age, which ended about 12,000 years ago (Kolbert 2014). Long before *homo sapiens* is threatened with extinction, we will add genocide to ecocide by fighting amongst ourselves for the control of the ever-shrinking parts of the planet that can sustain us. For future generations, anthropocide will thus be experienced as economic, political, and social collapse, as a reversal of what Norbert Elias (1939) referred to as the civilising process.

The IPCC (2022a) is, once again, clear on how this reversal will unfold, dividing the impacts of climate change into two categories: ecosystems and human systems. Human systems are, however, doubly vulnerable because they can be disrupted by climate change directly *and* indirectly through disruption to ecosystems. A very brief summary of the IPCC's predictions for 2041 to 2060 (*medium term*) is: water scarcity, challenges to food production, increased mortality, increased mental illness, and damage to urban and national infrastructures. The IPCC provides a further level of detail with the category of Complex, Compound, and Cascading Risks. The prediction here, again very briefly, is that climate change will continue to destabilise social, cultural, religious, political, and economic systems, exacerbating already-existing social inequalities and political polarisation. This is the crisis with which I am concerned, the slow apocalypse that has already been set in motion by the natural challenge of climate change and is being facilitated and aggravated by climate change as a cultural challenge. If this crisis is not strictly speaking anthropocidal, it is unquestionably genocidal and will radically alter humanity's mode of being, what Ludwig Wittgenstein (1953 §19: 11) called a 'form of life' and Raymond Williams (1958: 42) a 'structure of feeling.' Either humanity will fail to meet the cultural challenge, making the world a much more dangerous and divided place, or it will meet the cultural challenge by reconstructing civilisation as it currently exists. This reconstruction will, I suggest, involve a reconfiguration of the *oikeios*.

1.2 The Capitalocene

Jason Moore (2015: 2, emphasis in original) opens *Capitalism in the Web of Life: Ecology and the Accumulation of Capital* with this definition: 'Capitalism is not an economic system; it is not a social system it is *a way of organizing nature*.' He warns against the separation of capitalism or modernity from nature or ecology because the two have been inextricably linked in a relation of life-making since the long 16th century (1451–1648). As such, Moore (2015: 4; see also Wallerstein 1974) adopts a world-system approach he refers to as 'capitalist world-ecology.' Capitalist world-ecology is not an interaction of world-economy and world-ecology because 'world-economies *are* world-ecologies' (Moore 2015: 197, emphasis in original). He regards human exceptionalism, the view that human beings are independent of the spatiotemporal web of interspecies dependencies, as deeply misguided. Human agency has always been a part of and inextricably bound to nature, and Moore's inquiry begins with the relations between culture and nature, the co-production of human and nonhuman animal life, and the environments that maintain them. He (Moore 2015: 35) describes this life-making relation as the *oikeios*, which is 'a way of naming the creative, historical, and dialectical relation between, and also always within, human and extra-human natures.' Taking the *oikeios* rather than human agency or culture as his starting point, Moore argues that civilisations develop through nature rather than interacting with nature.

Capitalist world-ecology has been converting energy into capital in increasingly innovative and expansive ways since the coincidence of the Dutch agricultural revolution, Central European mining revolution, and Madeiran sugar–slave nexus in 1450. This year marks the beginning of the 'Capitalocene' (Age of Capital), in contrast with and in opposition to the 'Anthropocene' (Age of Man) (Moore 2015: 77). The Anthropocene is used to describe the geological epoch during which humanity has had a significant human impact on the Earth's ecosystems and climate and is usually dated to either the Industrial Revolution (1760–1840) or the First Agricultural Revolution (10,000 BCE). In Moore's assessment, the current epoch is characterised not by the impact of humanity as a species but by the impact of capitalist civilisation, which was inaugurated between the two revolutions. Moore (2015: 119–120) divides capitalist world-ecology into five historical periods as follows:

> (1) a Germanic-Iberian cycle (c.1451–1648), in which the expansionary phase turns to relative decline after the 1557 financial crisis; (2) a Dutch-led cycle (c.1560s–1740s), in which decline sets in after 1680; (3) a British-led cycle, c.1680s–1910s), with relative decline after 1873; (4) an American-led cycle (c.1870s–1980s), with relative decline after 1971; and (5) a neoliberal cycle (it could just as easily be called neo-mercantilist) that commenced in the 1970s.

The five cycles of accumulation and appropriation are both a cause and a consequence of the reorganisation of capitalist world-ecology. The capitalist epoch emerged from the crisis of the feudal epoch in the long 14th century (1290–1453), which was a consequence of the coincide of the inability of agriculture to sustain population growth, the Little Ice Age, the Black Death, and the Hundred Years' War. Moore (2015: 124, emphasis in original) contends that the crucial difference between modernity and pre-modernity, capitalism and feudalism, is the former's 'constant enlargement – and revolutionizing – of the geographies of *potential* accumulation and appropriation.'

Ceaseless accumulation and appropriation are facilitated by 'Cheap Natures, understood primarily as the "Four Cheaps" of labor-power, food, energy, and raw materials' (Moore 2015: 17). An increase in the cost of one or more of Cheap Food, Cheap Labour, Cheap Energy, or Cheap Raw Material is met with a new Cheap Nature strategy, which employs technological innovation, geographical expansion, or both to keep the price of production down. The early modern scientific revolutions were thus a crucial component of the Capitalocene, introducing a rationality based on 'mathematical abstraction and cartographic perspective' (Moore 2015: 207). The contribution of the Industrial Revolution was to transform capitalism into a planetary system, which differentiated the last three cycles from the first two in two ways: (1) the value relations of the Atlantic world achieved global hegemony and (2) accumulation and appropriation achieved unprecedented levels. The peak of the world-ecological surplus and, in consequence, of appropriation was at the height of the British cycle (1817–1870). The foundation of capitalist world-ecology is that the law of Cheap Natures integrates human and extra-human natures in the web of life. 'At the core of this law is the ongoing, radically expansive, and relentlessly innovative quest to turn the work/energy of the biosphere into capital (value-in-motion)' (Moore 2015: 14). As long as Cheap Natures can be kept cheap by neo/colonial conquests, agro-industrial revolutions, or scientific paradigm shifts, the capitalist epoch will continue, shifting from cycle to cycle in order to sustain its relentless transformation of energy into capital.

The continual challenge faced by capitalist world-ecology is that the demand for Cheap Natures exceeds the capacity of production systems to meet it. This causes the costs of production to rise, which in turn causes accumulation to falter. The challenge of the insatiable demand for Cheap Natures and the reorganisation of world-ecology to meet that demand is what drives one cycle of accumulation and appropriation into the next. '*Accumulation by appropriation*' is the set of coercive, cultural, and calculative processes by means of which capital gains access to non-commodified or minimally commodified natures for free or cheaply (Moore 2015: 95, emphasis in original). Every cycle of accumulation begins with co-produced ecological surplus, and the price of the Four Cheaps is co-determined by the combination of extra-human natures and intra-human power relations. As demand exceeds the capacity to meet it, the Four Cheaps become less cheap until they reach a tipping point where they start becoming more expensive, signalling the exhaustion of the accumulation regime and the development of a systemic crisis. These crises are resolved by the 'trinity' of neo/colonial conquests, agro-industrial revolutions, and scientific paradigm shifts, which occur at different moments during the crises (Moore 2015: 150). The crises are resolved when the three uneven moments converge, restoring the Four Cheaps. Moore uses the transition between the two North Atlantic cycles as examples, the period between 1790 and 1960 when changes from coal and steam to oil and internal combustion were combined with rising contributions of unpaid labour (especially that of women), and the expansion of first the British and then the American empires. The transitions between cycles of accumulation and appropriation suggest the significance of racism and sexism, both of which can be employed to restore the Four Cheaps. There is no more powerful justification of a social or political hierarchy in which those at the top accrue capital at the expense of cheap or free labour by those at the bottom than by claiming the hierarchy to be a product of nature rather than culture.

Neoliberal capitalist world-ecology has, however, exhausted Cheap Natures in an unprecedented way because it is the culmination of successive cycles in which the ecological surplus declines while capitalisation increases, undercutting the basis of accumulation by appropriation. This, according to Moore (2015: 87), is 'the end of Cheap Nature – and with it, the end of capitalism's free ride.' The problem is exacerbated by climate change, which is not merely a problem with the cost of one or even all of the Four Cheaps but a 'paradigm moment of the transition to negative-value' (Moore 2015: 276). After centuries of internalising the waste of capital, the biosphere is forcing capital to internalise biospheric shifts. Unlike previous systemic crises, climate change presents an insurmountable challenge to capitalist world-ecology by requiring production systems to internalise the cost of waste. The contemporary moment is thus one in which the unprecedented exhaustion of Cheap Natures is coincidental with an unprecedented rise in the cost of Cheap Natures. Moore (2015: 276) is clear that there 'is no conceivable way that capitalism can address climate change in any meaningful way.' Capitalist world-ecology is reliant on the co-production of Cheap Natures, and with Cheap Food, Cheap Labour, Cheap Energy, and Cheap Raw Material increasingly expensive, it can no longer be sustained. Climate change therefore 'poses a fundamental threat not only to humanity, but, more immediately and directly, to capitalism itself' (Moore 2015: 290).

Climate change is responsible for the transition to negative value, which Moore (2015: 277) defines as 'the accumulation of limits to capital in the web of life that are direct barriers to the restoration of the Four Cheaps.' Historically, the accumulation of limits to capitalist world-ecology was a potentiality that was never realised. The belated recognition of human-induced (or, more accurately, co-produced) climate change required capital to internalise biospheric shifts, which has – for the first time in the capitalist epoch – realised the potentiality and initiated an accumulation of negative value. In the neoliberal cycle, the same technological advances in production that create demands in excess of the supply of raw materials have also produced 'a *general law of overpollution*: the tendency to enclose and fill up waste frontiers faster than it can locate new ones' (Moore 2015: 280, emphasis in original). As Cheap Natures became increasingly expensive, they also became increasingly toxic, creating a world in which capital's toxification is ubiquitous, from microplastics in human blood to the Great Pacific garbage patch north of Hawaii. The exhaustion of Cheap Natures and the toxification of the *oikeios* are systemic crises not just for the neoliberal cycle but also for capitalist world-ecology, and Moore (2015: 305) holds that the latter is likely to have more of an impact than the former: 'The end of cheap garbage may loom larger than the end of cheap resources.' He is certain that the capitalist epoch will be replaced in the 21st century but not of what the next epoch will be. Capitalism is one among many ways of organising nature, and nature itself can neither be destroyed nor saved, only organised in different ways in reciprocal relationships with human civilisations. Those reconfigurations can, in Moore's (2015: 48) terms, be 'more or less emancipatory, more or less oppressive' from the standpoint of humanity-in-nature and nature-in-humanity.

1.3 Green Criminology

Ecocide is typically investigated within 'Green Criminology,' which was coined by Michael Lynch (1990: 11). As a field or subdiscipline within criminology, green criminology is usually understood as an umbrella term, that is a loose perspective (South 1998). This perspective is described by White (2022: 19) as being 'premised on the idea that the justice system needs to take environmental harm seriously.' Green criminology is thus essentially rather than accidentally zemiological, focusing on harm rather than crime. Zemiology is another umbrella term, the origin of which is often cited as Julia and Herman Schwendinger's (1970) 'Defenders of Order or Guardians of Human Rights?', in which they challenged criminologists to turn their attention to social harms such as war, racism, sexism, and classism. This challenge was met by Paddy Hillyard, Christina Pantazis, Steve Tombs, and Dave Gordon (2004), who inaugurated zemiology with *Beyond Criminology: Taking Harm Seriously*. Zemiology takes preventable harms as its subject, regardless of whether or not those harms have been criminalised. Green criminology is, in consequence, concerned with environmental crimes and harms, environmental laws, environmental regulation, and eco-justice, which explores the value of nonhuman animals, plants, and ecosystems. The first monograph published in green criminology is relatively recent, White's (2008) *Crimes Against Nature: Environmental Criminology and Ecological Justice*, and White remains the most prolific author in the field to date, with exemplary works on both theory (White 2022) and practice (White 2023). In the 21st century, green criminology expanded and diversified to include subfields and specialisms such as green cultural criminology (Brisman & South 2014), southern green criminology (Goyes 2019), and gendered green criminology (Milne et al. 2023).

Given that climate change is the 'biggest issue in the history of humankind', White (2018: 149) maintains that it demands nothing less than the prioritisation of research, policy, and practice by criminologists. His (White 2018: 149) call to action is worth quoting at length:

> For criminologists, the role responsibilities are clear. We need to engage ourselves as public intellectuals and in political action, to assume the mantle of stewards and guardians of the future, and to prioritise research, policy and practice around climate change themes. The biggest issue in the history of humankind demands nothing less than this. Intervention in relation to climate change involves drawing upon . . . a wide variety of sources (for example, cross-disciplinary, multijurisdictional, cross cultural) to investigate matters such as conflicts over resources, climate-induced migration, and the social effects of radical shifts in weather patterns. Information and data that is collated can be analysed and interpreted in the light of broad eco-global criminological considerations (for example, transgressions against humans, ecosystems and animals), as well as specific patterns of environmental victimisation involving particularly vulnerable groups (for example, women, children, disadvantaged groups, ethnic minorities). Steps can be taken to theorise the findings in relation to anthropogenic causes (for example, human responsibility for harm, specific perpetrators

and degrees of culpability). All of these have been identified herein as crucial tasks of Climate Change Criminology.

In other words, White wants to reorientate (all) criminology as climate change criminology, prioritising the mass harm of ecocide over and above all other harms and crimes. I agree completely and extend his appeal beyond criminology to all of the disciplines within the social sciences and humanities. I am not suggesting that only research aimed at the reduction or prevention of ecocide has value, but that ecocide must be dealt with first or the rest will cease to matter. Such a change would indeed require the complete reorientation of the social sciences and humanities, but it would be no greater than the change from higher education as a public good to higher education as a commercial enterprise, which was achieved in less than a decade in the United Kingdom (beginning in the 2012–2013 academic year).

White underpins this reorientation with five pillars: harm, eco-justice, causality, power, and intervention. Climate change criminology is zemiological because most of the current carbon emissions are lawful in spite of being social and environmental harms. It must adopt a 'global perspective that views the world as an interconnected whole', paying particular attention to the relationship between human beings and global ecology (White 2018: 146). Like both critical and mainstream criminology, climate change criminology should be focused on the causes of harm, specifically the causes of anthropogenic global warming and its diverse consequences. It is primarily concerned with political economy and the role of the powerful, who are responsible for both the material harms of climate change and their representation. Finally, climate change criminology promotes 'public engagement and social interventions that challenge the status quo', including those directed against the state (White 2018: 147). White argues that ecocide should be recognised as a mass harm – the most important mass harm – and a crime against humanity in order to enable liability and prosecution by the criminal justice system. His action plan for reorienting the justice system in response to ecocide (which also has five parts and is based on his work with Ronald Kramer) begins the section on social action with the rhetorical 'and symbolic construction of climate change as "crime,"' which draws attention to what I refer to as the cultural challenge of climate change (White & Kramer 2015: 392).

References

Brisman, A. (2012). The Cultural Silence of Climate Change Contrarianism. In: White, R. (ed.). *Climate Change from a Criminological Perspective*. New York: Springer Publishing, 41–70.

Brisman, A. & South, N. (2014). *Green Cultural Criminology: Constructions of Environmental Harm, Consumerism, and Resistance to Ecocide*. Abingdon: Routledge.

Children of Men (2006). Directed by Alfonso Cuarón. US: Universal Pictures.

Elias, N. (1939/2000). *The Civilizing Process: Sociogenetic and Psychogenetic Investigations*. Trans. E. Jephcott. Malden, MA: Blackwell Publishing.

Emanuel, K. (2018). *What We Know about Climate Change*. Updated ed. Cambridge: MIT Press.

Glavovic, B.C., Smith, T.F. & White, I. (2022). The Tragedy of Climate Change Science. *Climate and Development*, 14 (9), 829–833.

Goyes, D. (2019). *Southern Green Criminology: A Science to End Ecological Discrimination*. Leeds: Emerald Publishing.

Higgins, P. (2010/2015). *Eradicating Ecocide: Laws and Governance to Prevent the Destruction of Our Planet*. London: Shepheard-Walwyn.

Hillyard, P., Pantazis, C., Tombs, S. & Gordon, D. (eds.) (2004). *Beyond Criminology: Taking Harm Seriously*. London: Pluto Press.

Intergovernmental Panel on Climate Change (IPCC) (2021). *Climate Change 2021: The Physical Science Basis*. Available at: <www.ipcc.ch/report/sixth-assessment-report-working-group-i/>.

Intergovernmental Panel on Climate Change (IPCC) (2022a). *Climate Change 2022: Impacts, Adaptation and Vulnerability*. Available at: <www.ipcc.ch/report/sixth-assessment-report-working-group-ii/>.

Intergovernmental Panel on Climate Change (IPCC) (2022b). *Climate Change 2022: Mitigation of Climate Change*. Available at: <www.ipcc.ch/report/ar6/wg3/>.

Intergovernmental Panel on Climate Change (IPCC) (2023). *Climate Change 2023: Synthesis Report*. Available at: <www.ipcc.ch/report/ar6/syr/downloads/report/IPCC_AR6_SYR_FullVolume.pdf>.

Intergovernmental Panel on Climate Change (IPCC) (2024). IPCC Agrees on the Set of Scientific Reports for the Seventh Assessment Cycle. *Newsroom*. Available at: <www.ipcc.ch/2024/01/19/ipcc-60-ar7-work-programme/>.

Intergovernmental Panel on Climate Change (IPCC) (2025). *About the IPCC*. Available at: <www.ipcc.ch/about/>.

Kolbert, E. (2014). *The Sixth Extinction: An Unnatural History*. New York: Henry Holt & Company.

Lynch, M. (1990). The Greening of Criminology: A Perspective for the 1990s. *Critical Criminologist*, 2 (3), 3–12.

McGregor, R. (2023). *Literary Theory and Criminology*. Abingdon: Routledge.

Milne, E., Davies, P., Heydon, J., Peggs, K. & Wyatt, T. (2023). *Gendering Green Criminology*. Bristol: Bristol University Press.

Moore, J. (2015). *Capitalism in the Web of Life: Ecology and the Accumulation of Capital*. New York: Verso Books.

The Royal Society (2025). Is the Climate Warming? *Climate Change: Evidence and Causes*. Available at: <https://royalsociety.org/news-resources/projects/climate-change-evidence-causes/question-1/>.

Schwendinger, H. & Schwendinger, J. (1970). Defenders of Order or Guardians of Human Rights? *Issues in Criminology*, 5 (2), 123–157.

South, N. (1998). A Green Field for Criminology? A Proposal for a Perspective. *Theoretical Criminology*, 2 (2), 211–233.

Wallerstein, I. (1974). *The Modern World-System: Capitalist Agriculture and the Origins of the European World-Economy in the Sixteenth Century*. New York: Academic Press.

White, R. (2008). *Crimes against Nature: Environmental Criminology and Ecological Justice*. Uffculme: Willan Publishing.

White, R. (2018/2020). *Climate Change Criminology*. Bristol: Bristol University Press.

White, R. (2022). *Theorising Green Criminology: Selected Essays*. Abingdon: Routledge.

White, R. (2023). *Advanced Introduction to Applied Green Criminology*. Cheltenham: Edward Elgar Publishing.

White, R. & Kramer, R.C. (2015). Critical Criminology and the Struggle Against Climate Change Ecocide. *Critical Criminology: An International Journal*, 23 (4), 383–399.

Whyte, D. (2020). *Ecocide: Kill the Corporation before It Kills Us*. Manchester: Manchester University Press.

Williams, R. (1958/1960). *Culture and Society 1780–1950*. New York: Anchor Books.

Wittgenstein, L. (1953/2009). *Philosophical Investigations*. Trans. G.E.M. Anscombe, P.M.S. Hacker & J. Schulte. Chichester: Wiley-Blackwell.

2 Culture

I began Chapter 1 by stating my intention to demonstrate that cinematic fictions have the capacity to confront the climate challenge by shaping the desires of their audiences. I disclosed my argument for this core claim as being composed of three constituent claims, each of which is the subject of one of the three parts of this essay. Part I, The Problem, identifies the climate challenge as a cultural challenge, which was the subject of Chapter 1, and suggests three solutions to the cultural challenge in this chapter. My rationale is that if the challenge of climate change is a cultural as well as a natural challenge (or, more accurately speaking, *primarily* a cultural challenge), then cultural artefacts are likely to provide resources for and to meeting that challenge. This is, indeed, the foundation of cultural criticism, which is premised on the idea that expertise in cultural artefacts such as literary, cinematic, and narrative texts provides a means to the end of analysing and evaluating the cultures within which those texts are produced and consumed (McGregor 2024). If one accepts the cultural critical paradigm – at least until Chapter 3, in which I critique it – then there are three ways in which narratology, as my own area of expertise, might contribute to confronting the challenge of climate change, through: green cultural criminology, monomythic storytelling, and aesthetic education.

2.1 Green Cultural Criminology

In Chapter 1, I delineated green criminology as a field within criminology *premised on the idea that the justice system needs to take the environment seriously and focused on environmental crimes and harms, environmental laws, environmental regulation, and eco-justice*. Green criminology is one of several subdisciplines that developed during the expansion of the discipline in the last decade of the 20th century, but was relatively slow to gain momentum and remained without a published monograph until 2008. I noted that green criminology had itself expanded and diversified in the second decade of the 21st century and that one of the specialisms developed was green cultural criminology. Green cultural criminology was inaugurated by Avi Brisman and Nigel South (2014) with the publication of *Green Cultural*

DOI: 10.4324/9781003565826-3

Criminology: Constructions of Environmental Harm, Consumerism, and Resistance to Ecocide in 2014. The aim of this concise but inspired monograph is, in the authors' (Brisman & South 2025: 2, emphasis in original) words, 'to bring together *green criminology* and *cultural criminology*, and to identify points of overlap.' This synthesis has been so successful that the subfield has itself expanded and diversified, with subsequent authors focusing on the overlap between green cultural criminology and visual criminology (Natali & McClanahan 2017) and green cultural criminology and southern green criminology (Goyes et al. 2021). So far I have, however, said little about cultural criminology.

Like green criminology, cultural criminology was one of the subdisciplines to emerge from the expansion of criminology in the 1990s. Unlike green criminology, cultural criminology had a more directly sociological genealogy, following from the subcultural theories developed by Albert Cohen (1955) and David Matza (1964) and the work of Stuart Hall and his collaborators at first the University of Birmingham (Hall et al. 1978) and then the Open University (Hall, Evans & Nixon 1997) on media, representation, and meaning. Broadly construed, cultural criminology is a framework or paradigm that deploys theories, principles, and methods from a US-based tradition of anthropology and a UK-based tradition of media studies, which are more or less successfully integrated. The framework was pioneered by Jeff Ferrell (1996), with *Crimes of Style: Urban Graffiti and the Politics of Criminality*, and by Keith Hayward (2004), with *City Limits: Crime, Consumer Culture and the Urban Experience*. According to Ferrell, Hayward, and Jock Young (2015), cultural criminology is essentially about the interweaving of cultural forces with the practices of crime and its control and aims to understand crime as expressive; understand crime as a global public spectacle, mediated for consumption; and provide a critique of the politics of crime and criminal justice. Ferrell's (1995: 29) initial description of the relationship between representation and reality refers to *mirroring*, the study of 'not only images, but images of images, an infinite hall of mediated mirrors.'

Ferrell, Hayward, and Young (2015: 155) develop this conception, which is:

> a circulating cultural fluidity that challenges any certain distinction between an event and its representation, a mediated image and its effects, a criminal moment and its ongoing construction within collective meaning. Importantly, this looping process suggests for us something more than Baudrillard's postmodern hyper-reality, his sense of an 'unreality' defined only by media images and obfuscation. Quite the opposite: we mean to suggest a late-modern world in which the gritty, on-the-ground reality of crime, violence and everyday criminal justice is dangerously confounded with its own representation.

With regard to popular culture, cultural criminologists have studied its artefacts in relation to what Marxists refer to as *ideology*. Karl Marx (Marx &

Engels 1846) introduced a new denotation of ideology, as not just a systematic scheme of ideas or set of beliefs governing conduct, but the false understanding that the beliefs and attitudes of the dominant class reflect that class' direct access to reality. Louis Althusser (1969) argued that ideology represented an imaginary relationship between individuals and the reality of their existence within the capitalist mode of production. This relationship is perpetuated and exploited by the state in consequence of its commitment to that mode of production. Drawing on Althusser, Slavoj Žižek (1989: 15–16, emphasis in original) proposes a more robust delineation as '*a social reality whose very existence implies the non-knowledge of its participants as to its essence*'. Nicole Rafter (2006: 9) defines ideology in terms of the myths that shape social reality, where *myth* is, in turn:

> a descriptive term for the fundamental notions that people hold (usually without much conscious thought) about how the world is structured, what is valuable and unworthy, who is good and who is bad, and which kinds of actions are wrong or right.

In consequence and in spite of notable exceptions such as green cultural criminology, the cultural criminological attention to representation (whether nonfictional or fictional in character) has largely been concerned with misrepresentation, with crime news and crime drama misrepresenting the reality of crime and its control.

If one summarises green criminology as the study of environmental crime and harm and cultural criminology as the study of culture as a site of meaning-making and as a tool for intervention in the politics of crime control, then green cultural criminology is the study of culture as a site of meaning-making and as a tool for intervention in the politics of climate change. In *Green Cultural Criminology*, Brisman and South (2014) explore the significance of the ways in which environmental crime, consumerism, and green activism are represented by the media and in popular culture. They demonstrate the value of the cultural criminological lens, particularly but not exclusively its conceptions of deviance and spatiality, as a tool for intervention in the politics of climate change. Brisman and South conclude that the way in which environmental harms are represented is crucial to their relation to the criminal justice system, that the commodification of nature is one of the causes of harmful patterns of consumption, and that the contestation of space is a successful means of resisting and ultimately reducing environmental harm. Their second monograph, *Monstrous Nature and Representations of Environmental Harms: A Green Cultural Criminological Perspective* (Brisman & South 2025), narrows their focus from the media and popular culture to popular culture, with an emphasis on fictional representations, especially narrative fictions such as novels, feature films, and television series. Brisman and South's argument proceeds in two stages: first, that the representation of nature as one or more of monstrous, abject, or apocalyptic is the dominant mode of representation in contemporary narrative fiction and, second, that this representation is harmful in virtue of the way in which it shapes the experiences, attitudes, and beliefs of audiences. While there is a sense in which Brisman and South retain the cultural criminological concern with misrepresentation, they recognise that this is a contingent rather than necessary feature of popular cultural artefacts and, more importantly from my perspective, acknowledge that those artefacts represent rather than merely reflect reality. Green cultural criminology is thus compatible with – or perhaps even a new direction for – cultural criticism and provides me with a framework from within which to locate my inquiry.

2.2 Monomyth

In *The Hero with a Thousand Faces*, which was first published in 1949, Joseph Campbell (2008: 23) describes the 'monomyth' as a unification and magnification of three rites of passage: separation, initiation, and return. The monomyth can be described in a pair of sentences (Campbell 2008: 23):

> A hero ventures forth from the world of common day into a region of supernatural wonder. . . . : fabulous forces are there encountered and a decisive victory is won . . . : the hero comes back from this mysterious adventure with the power to bestow boons on his fellow man. . . .

The lion's share of *The Hero with a Thousand Faces* comprises dozens of examples of the monomyth drawn from historical and contemporary religions in each continent as well as the Big Five global religions. The majority of these religions, including the Big Five, are at best phallocentric and at worst misogynist, which is why I have not embellished the quotation with '*sic*' or amended 'man' to 'human.' Campbell maintains that *every* historical and contemporary belief system is an instantiation of the monomyth, and his argument is convincing in virtue of the extent of the empirical evidence he cites. Each of the three rites or stages is subdivided, and Campbell provides a more detailed description of the monomyth in terms of eight scenes or sections. These divide neatly into what is very likely the first definition of narrative or story in the Western tradition, Aristotle's (2000) representation with a beginning, a middle, and an end. The initiation, beginning, or first act opens with the call of adventure and closes with the confrontation with the guardian of the threshold (the first three scenes). The separation, middle, or second act opens with the arrival in the unfamiliar world and closes with the acquisition of the reward (scenes four and five). The return, end, or third act opens with a reversal of the journey and closes with the restoration of the world (the last three scenes). This is monomythic storytelling or monomythic narrative structure.

In 1985, ten years after the unprecedented commercial success of Steven Spielberg's *Jaws*, a Disney screenwriter named Christopher Vogler wrote a memorandum in which he identified *Jaws* and subsequent Hollywood blockbusters as exhibiting the same structure as Campbell's monomyth. In Vogler's (1992) memorandum, which was subsequently expanded to book length and published, he distilled the monomyth into a simple formula, a design consisting of 12 scenes divided into 5 acts. John Yorke (2013) provides an analysis of Vogler's model and argues that an overwhelming majority of commercially successful films have followed the model exactly, that is proceeded through each of the 12 scenes by means of 5 acts. There is simply something about this particular mode of storytelling – he suggests a complex combination of features in the final chapter of his book – that resonates with human beings, no matter what their spatiotemporal location. Yorke cites scores of examples of feature films and television series that follow the model, in a manner similar to Campbell in the original, but the match with *Jaws* is the closest, which is to be expected, given that it is widely acknowledged as the first 'event movie,' 'tent pole picture,' or 'summer blockbuster' (Shone 2004: 28 & 37).

Yorke (2013) superimposes the following template on the five-act structure: call to action, initial objective achieved, the point of no return, all is lost (or won), and resolution in victory (or defeat). This template is best explained using a concrete example, and I shall revisit my use of Francis Ford Coppola's *Apocalypse Now* (1979) with David Grčki in *An Epistemology of Criminological Cinema* (Grčki & McGregor 2024) as it is one of the greatest films of the 20th century and has an appropriately apocalyptic theme.[1] The call to action is what motivates the protagonist to change the routine of his or her life by pursuing a particular goal: after wating in Saigon, Captain Benjamin L. Willard (played by Martin Sheen) accepts a mission to execute Walter E. Kurtz (played by Marlon Brando), a colonel in the 5th Special Forces Group who has set himself up as the absolute ruler of a Highlander tribe in Cambodia. Relatively quickly, an initial objective is achieved: the United States Navy patrol boat that will take Willard to Kurtz is dropped into the fictional Nùng River by a United States Army air cavalry regiment. The point of no return occurs in the third act, and Yorke maintains that it is almost always timed to the precise halfway point in feature films: at exactly 88 minutes (of the 176 from opening to closing credits), Willard exercises his agency for the first time, executing a collateral civilian casualty. Up until this point, he has been a physical and psychological passenger on the boat, acquiescing to the Navy chief's will despite outranking him. The point of no return or turning point is the point at which the resolution of the film becomes inevitable and the story of *Apocalypse Now* is the story of Willard's transformation from an automaton of the agency of others to a fully autonomous agent.

The turning point is also the point in the narrative where there is a change in storytelling, from the story being driven by what the protagonist wants to the story being driven by what the protagonist needs, that is by the protagonist being forced to overcome a character flaw or vice and, in so doing, achieving some kind of psychological self-actualisation (this process – like the *all is lost* below – can be reversed in tragedies where the psychological transformation proceeds from virtue to vice). What Willard wants is to follow orders, accepting the missions he is given without question, but what he needs is to activate his own judgement, choosing how to complete his missions himself. As soon as he executes the wounded civilian, he is firmly set on the path to becoming a fully autonomous agent who will complete the Kurtz mission on his own terms and with formidable efficiency. At some stage during the fourth act, usually at its end, the story reaches a point where it either appears that all is lost and the protagonist is certain to fail to achieve his or her goal (if the story ends in victory) or where it appears that all is won and the protagonist is certain to succeed in achieving his or her goal (if the story ends in defeat). All appears lost as soon as Willard reaches Kurtz's compound, where he is immediately overpowered, taken prisoner, and tortured. Resolution can involve either victory or defeat and is either a reversal of the 'all is lost' situation or a reversal of the 'all is won' situation. This reversal is often subdivided into separate parts: the last ten minutes of *Apocalypse Now* is a carefully constructed physical and psychological climax in which the recently released Willard decides to execute Kurtz (166 minutes), infiltrates his quarters (167 minutes), hacks him to death with a machete (169 minutes), confronts the entire Highlander tribe (172 minutes), and then sets off downriver with the only surviving sailor (174 minutes). Ultimately, Willard follows his orders to the letter, but what is significant about his actions is that he has chosen to follow them and made his life meaningful in the exercise of his agency.

In monomythic terms: Willard (the hero) ventures forth from Saigon (the world of common day) into Cambodia (a region of supernatural wonder); Kurtz and his subjects (fabulous forces) are there encountered and a decisive victory is won; Willard (the hero) comes back from executing Kurtz (this mysterious adventure) with fully autonomous agency (the power to bestow boons on his fellow man). Mythic storytelling facilitates the communication of information in both an accessible and engaging manner. The information is accessible because we know what to expect (consciously or subconsciously) when we watch a monomythic cinematic work – because we have literally heard, read, or seen this story before – which makes it easier to follow the substantive details represented. The information is engaging because this particular way of telling a story seems to stimulate audience activity and Tom Shone (2004) characterises the blockbuster in terms of both audience participation (people talking about the film and buying related merchandise) and a relatively high proportion of repeat viewings. The monomyth thus appears to have the capacity to change attitudes towards climate change in a positive way.

2.3 Aesthetic Education

'Aesthetic education' is a vague and misleading term that has and is used to describe a variety of practices, from education in fine art and literature to the use of fine art and literature to enhance the teaching curriculum to a liberal arts education more generally. My use of *aesthetic education* in this essay is in its philosophical sense, which is the idea, hypothesis, or theory that the cultivation of aesthetic (or artistic) sensibility promotes both psychological (or ethical) and political (or social) harmony. As such, 'aesthetic education' is not only misleading but a misnomer because it actually refers to *a political education achieved by aesthetic means* (McGregor 2023, 2026). Aesthetic education was inaugurated by Johann Christoph Friedrich von Schiller (1795), the Württemberger polymath who wrote a series of letters in response to the Reign of Terror (1793–1794) in 1794. The letters were published as *Über die Ästhetische Erziehung des Menschen in einer Reihe von Briefen* (*On the Aesthetic Education of Man in a Series of Letters*), abbreviated to *Ästhetische Briefe* (*Aesthetic Letters*), in 1795, and made landmark contributions to the disciplines of art criticism and philosophy, establishing the first comprehensive theory of the significance of the relationship between artistic form and artistic content and the first comprehensive theory of aesthetic education. Schiller contended that aesthetic sensibility was an essential requirement for human flourishing, which was identical with the development of moral character, which was in turn a requirement for social cohesion at the national level.

The first point to note is that aesthetic education was conceived as an ideal solution to a concrete problem, the problem being the descent of the French Revolution (which Schiller admired) in to the Reign of Terror (which Schiller abhorred). In spite of acknowledging the aesthetic state as a utopian ideal, Schiller was convinced that the practical implementation of his theory would – in criminological terms – control, reduce, or prevent political violence (see McGregor 2026). My second point is that Schiller drew on his expertise as a philosopher and poet in establishing aesthetic education and, in consequence, drew on two distinct sets of predecessors, most notably Immanuel Kant and Johann Wolfgang von Goethe. In a similar manner, the tradition of aesthetic education he inaugurated has been sustained by a combination of poets and philosophers across the 19th, 20th, and 21st centuries. Walt Whitman and Matthew Arnold (1880) were particularly influential successors and developed the tradition by achieving fame as cultural critics as well as poets. The former is famous for what Jason Frank (2007) calls his *aesthetic democracy* and the latter for his claim that poetry was in the process of replacing religion. The emancipatory potential of the tradition was evinced in the first half of the 20th century in first the Harlem Renaissance in the United States and then *Négritude* in the French colonies (Locke 1925; Rabaka 2016). The critique of racial injustice received an intersectional twist in the United States in the work of bell hooks (1994, 1996) and other Black feminists in the second half of the 20th century. This emancipatory potential was not lost on Marxist critics and philosophers, who developed their own version of aesthetic education before, during, and after the Russian Revolution (1917–1923). Initiated by György Lukács, Walter

Benjamin, and Bertolt Brecht, the tradition was reinvigorated by the works of Raymond Williams (1958) and Jacques Rancière (1987).

In literary theory and criticism, what I am referring to as aesthetic education is usually known as the ethics of reading (which is also something of a misnomer, as aesthetic education is ultimately – and most importantly – about politics rather than ethics), after Joseph Hillis Miller's (1987) *The Ethics of Reading: Kant, De Man, Eliot, Trollope, James, and Benjamin*, which was published in 1987 and inaugurated a poststructural strand of ethical criticism that would draw on the work of Emmanuel Levinas (1961, 1974, 1982) and be exemplified in the work of Gayatri Chakravorty Spivak (1999, 2012, 2014). Two years later, Richard Rorty (1979, 1989) published *Contingency, Irony, and Solidarity*, which drew on his pragmatism and the philosophical tradition inaugurated by Georg Wilhelm Friedrich Hegel. This was followed by Martha Nussbaum's (1990) *Love's Knowledge: Essays on Philosophy and Literature* and Fredric Jameson's (1991) *Postmodernism, or, the Cultural Logic of Late Capitalism*. The former drew on the Aristotelian tradition of philosophy, was the first work in the new field of literary aesthetics, and was followed by *Poetic Justice: The Literary Imagination and Public Life* (Nussbaum 1995), which set out an explicit theory of aesthetic education. *Postmodernism, or, the Cultural Logic of Late Capitalism* was the first of a planned seven-volume series that Jameson (1991, 2002, 2005, 2007, 2013, 2019) called *The Poetics of Social Forms* and was intended to map the relations between aesthetic form and political economy when complete. Jameson's (1981) *Poetics* is itself a Marxist theory of aesthetic education and was pre-empted by his publication of *The Political Unconscious: Narrative as a Socially Symbolic Act* in 1981.

These four strands of aesthetic education – Hegelian, Marxist, Aristotelian, and Levinasian – constitute the tradition in its contemporary instantiation. The actual mechanism that connects aesthetic sensibility to psychological and political harmony differs in each strand, but all four retain Schiller's original conception that the means is indirect rather than direct. Aesthetic experience is an experience of play – between the sensory and the rational for Schiller – and the experience of play facilitates self-actualisation and human flourishing in contemporary terms. The synthesis of the physical and the cognitive in play is what provides the aesthetic with its political relevance, and Schiller (1795, II: 9) held that 'if man is ever to solve that problem of politics in practice he will have to approach it through the problem of the aesthetic.' All four of Rorty, Nussbaum, Jameson, and Spivak have focused on the novel as an implement of aesthetic education, and their central theses can be summarised as follows:

Rorty (Hegelian): novels can reveal the limitations of readers' beliefs and attitudes.

Jameson (Marxist): novels can provide new ways of thinking about the world in their form.

Nussbaum (Aristotelian):	novels can structure readers' imaginations as a play of particular and universal.
Spivak (Levinasian):	novels can enable readers to exercise their imaginations in an emancipatory way.

Although there is value in all four of the contemporary strands of aesthetic education, I shall focus on Jameson in this essay.[2] The reason for this focus is the significance he accords to allegory as one of, if not the most, complex kinds of narrative representation.

Notwithstanding Jameson's death in 2024, aged 90 and writing the final volume of the series, his *Poetics* is an unparalleled contribution to both literary theory and cultural criticism. He aimed to establish a comprehensive theory of the set of relations between historical and contemporary art forms and capitalist political economy, focusing on the six aesthetic modes of what Jason Moore (2015) calls the Capitalocene: myth (unfinished), allegory (Jameson 2019), realism (Jameson 2013), modernism (Jameson 2002, 2007), postmodernism (Jameson 1991), and utopianism (Jameson 2005). Jameson's critical practice is similarly unparallelled in its scope, analysing texts as diverse as Flemish Baroque painting, *opera seria*, European science fiction, postmodern architecture, experimental video, and American television for their insight into late capitalist culture. His theory of aesthetic education – which I summarised as *novels can provide new ways of thinking about the world in their form* but can be extrapolated to other types of narrative and non-narrative representations – is based on three interlocking concepts: postmodernism, allegory, and utopianism. More specifically, these concepts can be understood as what Mark Fisher (2022) refers to as capitalist realism (which I discuss in Chapter 5), genuine allegory (which I discuss in Chapter 3), and utopian desire (which I discuss in Chapter 8). To integrate two paraphrases: novels, which are both a site of compliance and resistance with and to, capitalist culture can provide utopian ways of thinking about the world in their form. My concern is of course with a film rather than a novel, but Alfonso Cuarón's *Children of Men* (2006) exhibits all three of these features as a self-reflective product of capitalist culture which is allegorical in form and embodies a utopian desire represented as an alternative to a dystopian present.

Notes

1 Three versions of the film have been released, and the analysis that follows is of *Apocalypse Now: Final Cut* (2019).

2 I discuss Nussbaum in *The Value of Literature* (McGregor 2016) and Nussbaum, Rorty, and Spivak in *Reducing Political Violence: Narrative, Critique, and Criminology* (McGregor 2026).

References

Althusser, L. (1969/2013). *On the Reproduction of Capitalism: Ideology and Ideological State Apparatuses*. Trans. G.M. Goshgarian. London: Verso Books.

Apocalypse Now (1979). Directed by Francis Ford Coppola. US: United Artists.

Apocalypse Now: Final Cut. (2019). Directed by Francis Ford Coppola. US: StudioCanal.

Aristotle (2000). *Poetics*. Trans. P. Murray & T.S. Dorsch. In: Murray, P. (ed.). *Classical Literary Criticism*. London: Penguin Books, 57–97.

Arnold, M. (1880/1925). The Study of Poetry. In: Arnold, M. *Essays in Criticism: Second Series*. London: Macmillan Publishers, 1–55.

Brisman, A. & South, N. (2014). *Green Cultural Criminology: Constructions of Environmental Harm, Consumerism, and Resistance to Ecocide*. Abingdon: Routledge.

Brisman, A. & South, N. (2025). *Monstrous Nature and Representations of Environmental Harms: A Green Cultural Criminological Perspective*. Philadelphia, PA: Temple University Press.

Campbell, J. (2008). *The Hero with a Thousand Faces*. 3rd ed. Novato, CA: New World Library.

Children of Men (2006). Directed by Alfonso Cuarón. US: Universal Pictures.

Cohen, A. (1955). *Delinquent Boys: The Culture of the Gang*. New York: Free Press.

Ferrell, J. (1995). Culture, Crime, and Cultural Criminology. *Journal of Criminal Justice and Popular Culture*, 3 (2), 25–42.

Ferrell, J. (1996). *Crimes of Style: Urban Graffiti and the Politics of Criminality*. Boston, MA: Northeastern University Press.

Ferrell, J., Hayward, K. & Young, J. (2015). *Cultural Criminology: An Invitation*. 2nd ed. London: Sage Publications.

Fisher, M. (2022). *Capitalist Realism: Is There No Alternative?* 2nd ed. Alresford: Zero Books.

Frank, J. (2007). Aesthetic Democracy: Walt Whitman and the Poetry of the People. *The Review of Politics*, 69 (3), 402–430.

Goyes, D.R., Abaibira, M.A., Baicué, P., Cuchimba, A., Ñeñetofe, D.T.R., Sollund, R., South, N. & Wyatt, T. (2021). Southern Green Cultural Criminology and Environmental Crime Prevention: Representations of Nature Within Four Colombian Indigenous Communities. *Critical Criminology: An International Journal*, 29 (3), 469–485.

Grčki, D. & McGregor, R. (2024). *An Epistemology of Criminological Cinema*. Abingdon: Routledge.

Hall, S., Critcher, C., Jefferson, T., Clarke, J. & Roberts, B. (1978). *Policing the Crisis: Mugging, the State, and Law and Order*. London: Macmillan Publishers.

Hall, S., Evans, J. & Nixon, S. (eds.) (1997). *Representation*. London: Sage Publications.

Hayward, K. (2004). *City Limits: Crime, Consumer Culture and the Urban Experience*. London: Glasshouse Press.

hooks, b. (1994). *Teaching to Transgress: Education as the Practice of Freedom*. New York: Routledge.
hooks, b. (1996). *Reel to Real: Race, Class and Sex at the Movies*. New York: Routledge.
Jameson, F. (1981/2002). *The Political Unconscious: Narrative as a Socially Symbolic Act*. New York: Routledge.
Jameson, F. (1991). *Postmodernism, or, the Cultural Logic of Late Capitalism*. Durham, NC: Duke University Press.
Jameson, F. (2002). *A Singular Modernity: Essay on the Ontology of the Present*. London: Verso Books.
Jameson, F. (2005). *Archaeologies of the Future: The Desire Called Utopia and Other Science Fictions*. London: Verso Books.
Jameson, F. (2007). *The Modernist Papers*. London: Verso Books.
Jameson, F. (2013). *The Antimonies of Realism*. London: Verso Books.
Jameson, F. (2019). *Allegory and Ideology*. London: Verso Books.
Levinas, E. (1961/1969). *Totality and Infinity: An Essay on Exteriority*. Trans. A. Lingis. Pittsburgh, PA: Duquesne University Press.
Levinas, E. (1974/1998). *Otherwise than Being, or, Beyond Essence*. Trans. A. Lingus. Pittsburgh, PA: Duquesne University Press.
Levinas, E. (1982/1985). *Ethics and Infinity: Conversations with Philippe Nemo*. Trans. R.A. Cohen. Pittsburgh, PA: Duquesne University Press.
Locke, A.L. (1925). *The New Negro: An Interpretation*. New York: Albert & Charles Boni Publishing Company.
Marx, K. & Engels, F. (1846/2000). The German Ideology. Trans. anonymous. In: Marx, K. (ed.). *Karl Marx: Selected Writings*. 2nd ed. Oxford: Oxford University Press, 175–208.
Matza, D. (1964). *Delinquency and Drift*. New York: John Wiley & Sons.
McGregor, R. (2016). *The Value of Literature*. London: Rowman & Littlefield International.
McGregor, R. (2023). Introduction: Aesthetic Education through Narrative Art. *Journal of Aesthetic Education Special Issue: Aesthetic Education through Narrative Art*, 57 (3), 1–11.
McGregor, R. (2024). Making Cultural Criticism Matter. *International Journal for the Semiotics of Law*. DOI: 10.1007/s11196-024-10192-6.
McGregor, R. (2026). *Reducing Political Violence: Narrative, Critique, and Criminology*. Bristol: Bristol University Press.
Miller, J.H. (1987). *The Ethics of Reading: Kant, De Man, Eliot, Trollope, James, and Benjamin*. New York: Columbia University Press.
Moore, J. (2015). *Capitalism in the Web of Life: Ecology and the Accumulation of Capital*. New York: Verso Books.
Natali, L. & McClanahan, B. (2017). Perceiving and Communicating Environmental Contamination and Change: Towards a Green Cultural Criminology with Images. *Critical Criminology: An International Journal*, 25 (2), 199–214.
Nussbaum, M. (1990). *Love's Knowledge: Essays on Philosophy and Literature*. New York: Oxford University Press.
Nussbaum, M. (1995). *Poetic Justice: The Literary Imagination and Public Life*. Boston, MA: Beacon Press.

Rabaka, R. (2016). *The Negritude Movement: W.E.B. Du Bois, Leon Damas, Aime Cesaire, Leopold Senghor, Frantz Fanon, and the Evolution of an Insurgent Idea*. Lanham, MD: Rowman & Littlefield.

Rafter, N. (2006). *Shots in the Mirror: Crime Films and Society*. 2nd ed. New York: Oxford University Press.

Rancière, J. (1987/1991). *The Ignorant Schoolmaster: Five Lessons in Intellectual Emancipation*. Trans. K. Ross. Redwood City, CA: Stanford University Press.

Rorty, R. (1979). *Philosophy and the Mirror of Nature*. Princeton, NJ: Princeton University Press.

Rorty, R. (1989). *Contingency, Irony, and Solidarity*. New York: Cambridge University Press.

Schiller, J.C.F.v. (1795/1967). *On the Aesthetic Education of Man: In a Series of Letters*. Trans. E.M. Wilkinson & L.A. Willoughby. London: Oxford University Press.

Shone, T. (2004). *Blockbuster: How Hollywood Learned to Stop Worrying and Love the Summer*. London: Simon & Schuster.

Spivak, G.C. (1999). *A Critique of Postcolonial Reason: Toward a History of the Vanishing Present*. Cambridge, MA: Harvard University Press.

Spivak, G.C. (2012). *An Aesthetic Education in the Era of Globalization*. Cambridge, MA: Harvard University Press.

Spivak, G.C. (2014). *Readings*. London: Seagull Books.

Vogler, C. (1992). *The Writer's Journey: Mythic Structure for Storytellers and Screenwriters*. Studio City, CA: Michael Wiese Productions.

Williams, R. (1958). *Culture and Society*. New York: Columbia University Press.

Yorke, J. (2013). *Into the Woods: How Stories Work and Why We Tell Them*. London: Penguin Books.

Žižek, S. (1989/2008). *The Sublime Object of Ideology*. London: Verso Books.

3 Critique

After setting out the climate challenge as a cultural challenge in Chapter 1, I suggested three ways in which narrative as a cultural artefact might provide a resource for confronting this challenge in Chapter 2. I noted that the practice of cultural criticism is founded on the premise that expertise in cultural artefacts provides a means to the end of analysing and evaluating the cultures within which those artefacts are produced and consumed. The idea I am developing in this essay is that narrative representation can be a resource for the analysis and evaluation of attitudes to and desires about climate change. Many social scientists reject this conception, contending that expertise in narrative authorises one to analyse and evaluate nothing more than narrative and that if those narratives are fictional, then the expertise is divorced from the reality in which they are experienced. If one wants to acquire expertise in attitudes to and desires about climate change, one needs to study those attitudes and desires directly, using the tried and tested methods, principles, and theories of social scientific practice. This scepticism about the epistemic value of cultural artefacts in general and narrative fictions in particular underpins a serious objection to each of my three proposals in Chapter 2: green cultural criminology, monomythic storytelling, and aesthetic education.

3.1 Fiction?

Jon Frauley (2010: 55) criticises the cultural criminological framework on the basis that it cannot reach beyond the 'textual image' – that is beyond the representation – and can only provide knowledge of the beliefs, attitudes, and desires of its creator and audience. The framework is thus primarily concerned with the production and reception of representations rather than with the reality represented. Notwithstanding, Frauley (2010, 2021a, 2021b) eschews *representation* to describe the relationship between text and context, referring instead to *reflection*. Specifically, the text is either a '*reflection of* reality' or '*reflects on* reality' and neither of these types of reflection provide the audience with knowledge of the causes or control of crime or harm (Frauley 2021b: 119 & 120, emphasis in original). This recalls Jeff Ferrell's

DOI: 10.4324/9781003565826-4

(1995) initial description of the same relationship: *we study not only images, but images of images, an infinite hall of mediated mirrors*. Although the cultural criminological framework was the first to take fiction seriously, its treatment of fictional representations has been almost exclusively negative, as a distorted reflection or a misrepresentation of reality. With regard to popular culture, cultural criminologists have attended to what Michelle Brown (2004: 220) calls the 'sizable gap in the field between what criminologists know about crime and what everyone else assumes.' The distinction is between the epistemological expertise of the former and the ideological ignorance of the latter. As discussed in Chapter 2, Nicole Rafter (2006) describes ideology in terms of the myths that shape social reality.

Writing at the turn of the century, David Garland (2001) argues that the mass media institutionalised a specific view of crime and punishment by means of tabloid news and television fiction, with a culture of control replacing a culture of correctionalism in the United States and the United Kingdom in a little over two decades. Gray Cavender (2004) claims that Garland understates the influence of the media, on the basis of televised documentary and drama producing a mutually reinforcing flow of information which fashioned the public endorsement of punitive policies. Neither Garland nor Cavender holds that the mass media caused high crime rates to become an organising principle of social reality in the 1970s but rather that cultural representations of rising crime both evinced and perpetuated the emergence of a widespread fear of crime. This reciprocal relationship provides a concrete example of the *infinite hall of mediated mirrors*. The rising crime rate in the United Kingdom from the mid-1950s and in the United States from the mid-1960s was reflected in the rising popularity of police procedural fiction. The combination of crime fiction and crime news shaped what Garland (2001: 163) calls the 'crime complex of late modernity', which was in turn reflected in the change in crime fiction in the 1970s, with more dangerous criminals requiring more ruthless police officers to protect the public. Cultural change in a constant direction over an extended period is more accurately characterised by Ferrell, Keith Hayward, and Jock Young's (2015: 155) *cultural looping*, where the 'mediated nature of contemporary culture not only carries along the meaning of crime and criminality; it circles back to amplify, distort and define the experience of crime and criminality itself.' Garland describes the changes in policy and culture from the 1960s to the 1990s as movement from an ideal of social progress to an ideal of zero tolerance, distorted and amplified through repeated looping to justify a security state built on the twin pillars of market discipline and moral discipline. This is the sense in which Ferrell, Hayward, and Young (2015) are sceptical of both documentary and dramatic representations, as creating a social reality where *the gritty, on-the-ground reality of crime, violence, and everyday criminal justice is dangerously confounded with its own representation.*

Frauley identifies Rafter as one of the few criminologists who have attempted to move beyond the limitations of the cultural criminological framework, although he does not believe that she explores the value of criminological fiction in sufficient detail. Rafter (2006) recognises four kinds of knowledge embedded in the fictional reality of cinematic representations (her research on criminological fiction has been restricted to feature films). Frauley (2010: 45, emphasis in original) regards the first two of these as being compatible with the cultural criminological framework: 'knowledge of *popular views* on the nature and causes of crime and criminality' and embodying a '*popularised version of criminological knowledge* that informs lay and populist views.' The creation of films thus draws on popular views about the causes and control of crime and harm, which are embedded in the patterns of meaning constitutive of the cinematic narrative. Third (Frauley 2010: 45, emphasis in original), films embody 'knowledge of *how ideology is connected to the power* to shape and mould popular knowledge as well as criminological knowledge.' This is an epistemic step beyond the cultural criminological framework as cultural criminologists typically regard crime films as evincing and perpetuating the relationship between ideology and power rather than reflecting on it. Finally, 'films offer *speculative knowledge* about the nature of crime in our society, in future and also past societies' (Frauley 2010: 45, emphasis in original). The denotation of 'speculative knowledge' is not entirely clear, but Frauley refers to its reception by audiences as potentially indicative of confidence in public policy and academic criminology and I take it to mean the exploration of the relationships between private and public, agency and structure, and individuals and types, that is the *criminological imagination.*

A second attempt to make a criminological move from representation to reality takes a different approach and seems to have been largely ignored by criminologists interested in fiction. While Cavender and Nancy Jurik (2012) mention neither cultural criminology nor the criminological imagination in *Justice Provocateur: Jane Tennison and Policing in* Prime Suspect, they are clearly conducting their research within the cultural criminological framework and clearly concerned with the criminological imagination. The criminological imagination draws on C. Wright Mills' (1959) conception of the sociological imagination and has been deployed by Frauley (2010) and Martin Glynn (2021) as a matrix within which to explore fiction and art for their criminological values. Cavender and Jurik (2012: 35) claim that their 'model directs attention to how much a cultural product portrays the structural context of individual problems, that is, to what extent it connects personal troubles with public issues.' They establish a model of progressive moral fiction, a genre characterised by insight into the experiences of the marginalised, the location of these experiences in a social context, the revelation of fissures in the ruling apparatus, and glimpses of hope to challenge injustice – all of which are represented from the point of view of a *justice provocateur*, an agent for positive change. The examination of ITV's *Prime Suspect* television series in these terms provides evidence of the model's value as a tool for both the analysis of narrative fiction and the exploration of social justice issues with students. Note, however, that as with Rafter and Frauley's own respective engagements with film, the representation does not – and is not intended to – reach the reality. The primary criminological value of *Prime Suspect* is pedagogical rather than epistemic, for providing an accessible and captivating way into the complexity of social and criminal justice rather than providing knowledge of the social and criminal justice that constitutes the context of the televisual text.

Simply put, the challenge to my argument in this essay is that if green cultural criminology emerged from cultural criminology, as discussed in Chapter 2 and acknowledged on multiple occasions by Avi Brisman and Nigel South (2014, 2025), then how can I justify moving from the representation (a cinematic fiction) to the reality beyond (the climate crisis)? This essay is an essay in green cultural criminology, not film criticism (or, more accurately, *primarily* an essay in green cultural criminology), and founded on the cultural critical premise that a cultural artefact can provide accurate and useful knowledge of the reality in which the artefact is produced and consumed, not merely knowledge of its own production and consumption. Strictly speaking, reflecting and looping do provide the knowledge of reality, but that knowledge is limited to the distortion and amplification characteristic of ideology, myth, and assumption. If one hones this challenge down further to expose its sharpest edge, it is essentially one of fictionality: how can Alfonso Cuarón's *Children of Men* (2006), set in a fictional future whose predictions for 2027 will almost certainly be revealed as inaccurate, communicate something that will be of practical use to criminologists and other social scientists? Not only is the idea counter-intuitive, but it also contradicts the core thesis of the cultural criminological framework from which green cultural criminology emerged.

3.2 Myth?

Joseph Campbell (2008) closes *The Hero with a Thousand Faces* with a reflection on the impact of first modernity and then globalisation on the monomyth. Given that the book was first published in 1949, he is remarkably prescient and inclusive, in spite of occasionally using language that became either dated or offensive by the time the third edition was published. Campbell fully anticipates the development of late modernity and what Francis Fukuyama (1992: xxi) would call 'the end of history' in *The End of History and the Last Man*. I discuss Fukuyama in Part II, but an overview of his thesis is that the collapse of the Iron Curtain in 1989 paved the way for a post-historical, post-political utopia in which liberalism and capitalism extend a benign hegemony over the Earth's nations courtesy of globalisation. Fukuyama was correct in predicting the disappearance of alternatives to capitalism in the early 1990s, but he failed to recognise the mass harms of the neoliberal cycle of capitalism, which has been characterised by market fundamentalism, hyper-individualism, and gross inequality. Here is what Campbell had to say in 1949 (2008: 334; the passage appears in all three editions):

> The problem of mankind today, therefore, is precisely the opposite to that of men in the comparatively stable periods of those great co-ordinating mythologies which now are known as lies. Then all meaning was in the group, in the great anonymous forms, none in the self-expressive individual; today no meaning is in the group – none in the world: all is in the individual. But there the meaning is absolutely unconscious. One does not know toward what one moves. One does not know by what one is propelled.
>
> The lines of communication between the conscious and the unconscious zones of the human psyche have all been cut, and we have been split in two.

Campbell is concerned about the survival of the monomyth in modernity in an age when meaning is located exclusively in the individual rather than the group – which is an apt summary of the semantics of late capitalist civilisation. What he failed to anticipate is the appropriation of the monomyth by Hollywood and its reconfiguration as the blockbuster, described by Christopher Vogler (1992) and Tom Shone (2004) and discussed in Chapter 2. As a myth created for a global rather than local audience, the Hollywood blockbuster cannot afford to integrate its structure with the meanings and values of a particular mythology, religion, or culture. The monomyth becomes, instead, a self-conscious product of the late capitalist imperative to maximise profit, seeking universal meanings and values that will appeal to all audiences but whose lack of specificity very often creates a narrative drained of any real significance. Hollywood has, in effect, pared the monomyth down to its most basic – banal, even – form, the hero's journey. Not even *the* hero's journey (the mythological, religious, or cultural hero), but *a* hero's journey. What hero? Anyone will do, from aquaphobic policemen (Brody in *Jaws*) to homicidal soldiers (Willard in *Apocalypse Now*) to deceitful matriarchs (Addy in *Us*; see Grčki & McGregor 2024).

The monomyth has, in consequence, become something of what Gayatri Chakravorty Spivak (2012: ix) refers to as a 'double bind'. Spivak deploys this concept ('matrix' may be more accurate) in several contexts, including her poststructural theory of aesthetic education, which is focused on the relationship between the aesthetic and the political in the era of globalisation. She draws attention to the paradox globalisation presents for aesthetic education: on the one hand, it enables what she refers to as *robust reading practices* (the foundation of her aesthetic education) to be established on an unprecedented scale, but on the other hand it forces robust reading practices (and the humanities more generally) into a profit-making model of education that reduces their value to the economic. In a similar manner, monomythic storytelling is now aimed at a global rather than local audience, but globalisation has deprived it of what matters most, the values that the hero shares with a particular mythology, religion, or culture. In both cases, what initially appears to be an opportunity is revealed to be a drain on the meaning and the value of the practice in question. Spivak's view is that the double bind is simultaneously an opportunity and a challenge, which exist in a reciprocal relation where each is dependent on the other, and I return to this idea in Chapter 7.

With the monomyth stripped back to its most meaningless and valueless instantiation – *monomyth-lite*, a hero's journey – it becomes difficult if not impossible to mobilise it to meet climate change as a cultural challenge. If the monomyth is nothing more than a hero's journey, then it propagates the individualism at the heart of neoliberal ideology (an ideology also described as post-social and post-political). There may be more than one hero – a pair, trio, or even nine heroes in J.R.R. Tolkien's *The Lord of the Rings* trilogy and their cinematic adaptations, but the monomyth as a narrative structure is only effective when there are a limited number of protagonists. Nine, for example, is an unusually large number and I have previously argued that even HBO's *Game of Thrones*, a television series that runs for over 70 hours, has only 8 protagonists. I (McGregor 2023: 136) actually refer to these eight as 'leading characters' and suggest that they can be reduced to three protagonists, one from each of the Houses of Lannister, Stark, and Targaryen. Climate change is a global threat to human and nonhuman animal existence. As such, effective solutions are likely to involve millions of people. If Jason Moore's (2015) claim that climate change is a direct consequence of the Capitalocene is accurate, then effective solutions are likely to involve a transformation of humanity's entire form of life or structure of feeling (most obviously, our relationship with nonhuman life, animal, and other), in which case those solutions would involve the entire species. How such a transformation could be communicated by means of the monomyth is unclear. The issue is not that directors and creative collectives are lacking in imagination, a deficiency that might be amended in the future, but that there is something inimical to confronting the climate crisis in monomythic storytelling.

The objection to my argument from myth or, more specifically and more accurately, from the monomyth is that if *Children of Men* (2006) follows Hollywood convention, then it utilises a monomythic structure that provides a misleading account of how the challenge of the climate crisis can be met. The monomyth is, as the title of Campbell's volume suggests, a hero's journey, a narrative in which an individual or multiple individuals overcome a challenge either in isolation or in temporary partnership. Climate (nature) does not respect national (cultural) borders, and climate change is not a challenge that can be met by arrogant billionaires on their own, by nations in isolation, or even by multiple nations temporarily united in coalitions. Climate change is, by definition, a global problem that requires a global response. In

Chapter 1, I delineated the climate challenge as a cultural challenge, and the pervasiveness of the monomyth in human storytelling indicates that at least one aspect of this cultural challenge is a narrative challenge. The monomyth is a narrative form that seems, at best, incompatible with the thematic content of climate change and, at worst, contradictory, a case of form undermining and deconstructing content. In other words, humanity needs not only new stories but also a new way of telling stories if those stories are to help us make sense of climate change.

3.3 Allegory?

Narrative form, which I discuss in Chapter 4, is basically the way in which a story is told – a map for the progress from its beginning to its middle to its end. This prototypical definition of a story or narrative is from Aristotle (2000), as noted in Chapter 2, and remains as accurate today as it was two and a half millennia ago. The most basic narrative imaginable requires an agent and two events. Not one event, but two, with the second occurring either after or before the first. This is because a narrative is, definitively, the representation of a *sequence of events*. A sequence is necessarily rather than contingently *temporal*: one thing happens and then another thing happens. I type this sentence *and then* I take a sip of coffee. Narrative represents time and is experienced over time. The sequence of events may be represented non-chronologically, when narratives jump backwards, forwards, or both in time or in a manner as complicated as, for example, Christopher Nolan's *Memento* (see McGregor 2014), but this does not alter the crucial character of narrative as temporal. One of the most complex narrative forms, no less because of its distinctive temporality, is allegory, and my interpretation and appreciation of *Children of Men* in Part II will be as an allegory.

Allegory is a convention in representation that combines duality with duplicity. The term is derived from the Greek word *allegoreo*, which means speaking otherwise, that is pretending to speak about one thing while actually speaking about another thing (Spivak 1999). Allegories describe or depict one subject in the guise of another subject, deploying the latter as a symbol, emblem, or personification of the former. Representation in allegories is twofold because the characters, settings, and actions in an allegory not only represent real or imagined people, places, and events but also represent something other, which may be a set of ideas, concepts, or abstractions or a second set of people, places, and events that are concealed, hidden, or secreted within the folds of the representation. Allegories can be verbal, visual, or a combination of the two and are not restricted to narrative representation, but also instantiated in painting and sculpture, which suggests a connection to the spatial to which I return below. Fredric Jameson (2019) is critical of this traditional model of allegory because it divides a narrative into two distinct lines with separate, hierarchical meanings. He (Jameson 2019: 10, emphasis in original) distinguishes twofold allegory from the *genuine allegory* in which he is interested as follows:

> [G]enuine allegory does not seek the meaning of a work, but rather functions to reveal its structure of multiple meanings, and thereby to modify the very meaning of the word *meaning*. It is indeed part of the contemporary critique of metaphysics (and of humanism along with it) to denounce the conception of nature as meaningful: an affirmation not merely of a meaningful system at work in the natural world, but also of a human nature as well, one which virtually by definition is normative. It will be clear, then, that this naturalization of both meaning and metaphor alike is the function of the symbol, as opposed to the allegorical structure.

Genuine allegory does not impose a meaning – or even a pair of meanings – on a representation but functions so as to reveal the structure of multiple meanings within the representation. Jameson maintains that this structure is fourfold.

The four levels at which meaning functions are the literal, the symbolic, the existential, and the anthropic. The *literal* level is simply the meaning of the represented sequence of events in the narrative, which are either real, imagined, or some combination of the two. The *symbolic* level is the hidden or secret meaning of the represented sequence of events. Jameson (2019: xvi) employs six different terms to describe the third level, which can be understood as the ethical meaning of the narrative representation, and I shall use *existential*, derived from his 'existential experience', as it captures both of the senses of ethics with which he is concerned, individual desire and the construction of subjectivity. Jameson (1981) describes the fourth and final level, which can be understood as the political meaning of the narrative representation, in terms of the Last Judgement of the Abrahamic and Zoroastrian eschatologies and his own conception of the political unconscious. The *anthropic* level is concerned with the 'collective and political narrative always latent in our conceptions of our own personal destinies' (Jameson 2019: xvii). The construction of subjectivity is embedded in a narrative of human achievement and Jameson's interest is in this collective ideology construed in terms of the species. He (Jameson 2019: xv) maintains that these four levels 'exhaust the various terrains on which ideology must perform its work' and summarises the levels at which meaning functions as: textual object, interpretive code, individual desire, and collective ideology.

The innovation in Jameson's model of allegory is not just that he replaces two levels of meaning with four and thus penetrates to layers of significance that the traditional model omits, but that fourfold allegories are *thick* narratives. In a thick narrative, the levels of meaning combine to create meaning in excess of the representational capacity of the text and this extra-representational meaning cannot be explained by reducing it to its constitutive levels. The thickness of fourfold allegory exceeds the representational capacity of narrative by activating an event in the reading (or viewing) process. Jameson (2019: 263) describes a 'narrative event' that is structural rather than representational, with the allegory revealing the complexity of its architecture of multiple meanings. The narrative is event is spatial rather than temporal in character (Jameson 2019: 234):

> This is the point at which to conclude that the spatial anomaly is not an aberration in the practice of allegory but rather constitutes the moment in which the heterogeneity of the allegorical levels breaks through and makes its presence felt. The levels are not a collection of complete narratives superimposed upon one another. Rather they come at reality in an utterly different way, by a jarring and sometimes dissonant differentiation of their various dimensions.

The intrusion of the spatial dimension of genuine allegory into the temporal dimension of narrative representation is what creates the narrative thickness and is the source of the extra-representational capacity of allegorical meaning. This spatial anomaly constitutes 'a reading experience utterly unique and quite distinct from either conventional narrative or pure lyric' (Jameson 2019: 263).

The spatial anomaly is consistent with Jameson's (1991) claim about postmodernism as an aesthetic mode that is essentially rather than accidentally atemporal and, in consequence, spatial in character. Spatiality is also why Jameson (2019: 117, 308, 347) maintains that allegory is particularly useful for shedding light on the complexity of 'our own moment of late capitalism', 'late capitalist globalization', and 'modern social life', revealing barely perceptible connections among the dimensions of late modern life. Genuine allegory provides an 'allegorical staging' of the complexity of social reality by means of the narrative event (Jameson 2019: 117). Allegorical narratives hold up both a mirror and a microscope to everyday life and can thereby illuminate causal relations that might otherwise remain unnoticed. As such, interpreting and appreciating *Children of Men* as an allegory seem eminently suited to my intention to employ it as part of some kind of programme of aesthetic education, as discussed in Chapter 2. The problem for my argument is that I have already described the film as an instantiation of the monomyth, a fundamentally temporal mode of representation. If, however, the value of allegory in revealing connections among the dimensions of late modern life is a function of its spatial character, then it seems likely that a narrative analysis or evaluation will diminish or even undermine its explanatory power. Jameson describes the spatial as *breaking through* and opposed to the temporal so one or the other must take precedence. Either, it seems, *Children of Men* is potentially useful *qua* monomyth (temporal) or *qua* allegory (spatial) – but not both.

References

Aristotle (2000). Poetics. Trans. P. Murray & T.S. Dorsch. In: Murray, P. (ed.). *Classical Literary Criticism*. London: Penguin Books, 57–97.

Brisman, A. & South, N. (2014). *Green Cultural Criminology: Constructions of Environmental Harm, Consumerism, and Resistance to Ecocide*. Abingdon: Routledge.

Brisman, A. & South, N. (2025). *Monstrous Nature and Representations of Environmental Harms: A Green Cultural Criminological Perspective*. Philadelphia, PA: Temple University Press.

Brown, M. (2004). Crime Fiction and Criminology. *Criminal Justice Review*, 29 (1), 206–220.

Campbell, J. (2008). *The Hero with a Thousand Faces*. 3rd ed. Novato, CA: New World Library.

Cavender, G. (2004). Media and crime policy: A reconsideration of David Garland's *The culture of control. Punishment & Society*, 6 (3), 335–348.

Cavender, J. & Jurik, N.C. (2012). *Justice Provocateur: Jane Tennison and Policing in Prime Suspect*. Champaign, IL: Illinois University Press.

Children of Men (2006). Directed by Alfonso Cuarón. US: Universal Pictures.

Ferrell, J. (1995). Culture, Crime, and Cultural Criminology. *Journal of Criminal Justice and Popular Culture*, 3 (2), 25–42.

Ferrell, J., Hayward, K. & Young, J. (2015). *Cultural Criminology: An Invitation*. 2nd ed. London: Sage Publications.

Frauley, J. (2010). *Criminology, Deviance, and the Silver Screen: The Fictional Reality and the Criminological Imagination*. New York: Palgrave Macmillan.

Frauley, J. (2021a). The Poverty of the Comparative Orthodoxy: Cultural Criminology, Perspectival Realism and Conceptual Variation. *International Journal of Law, Crime and Justice*, 66 (September), 1–12.

Frauley, J. (2021b). Fictional Realities and Criminology: Apprehending Social Reality Through Narrative Fiction. *Journal of Theoretical and Philosophical Criminology Special Edition: A Criminology of Narrative Fiction*, 13 (November), 111–125.

Fukuyama, F. (1992). *The End of History and the Last Man*. London: Penguin Books.

Garland, D. (2001). *The Culture of Control: Crime and Social Order in Contemporary Society*. Oxford: Oxford University Press.

Glynn, M. (2021). *Reimagining Black Art and Criminology: A New Criminological Imagination*. Bristol: Bristol University Press.

Grčki, D. & McGregor, R. (2024). *An Epistemology of Criminological Cinema*. Abingdon: Routledge.

Jameson, F. (1981/2002). *The Political Unconscious: Narrative as a Socially Symbolic Act*. New York: Routledge.

Jameson, F. (1991). *Postmodernism, or, the Cultural Logic of Late Capitalism*. Durham, NC: Duke University Press.

Jameson, F. (2019). *Allegory and Ideology*. London: Verso Books.

McGregor, R. (2014). Cinematic Philosophy: Experiential Affirmation in *Memento. The Journal of Aesthetics and Art Criticism*, 72 (1), 57–66.

McGregor, R. (2023). *Literary Theory and Criminology*. Abingdon: Routledge.

Mills, C.W. (1959). *The Sociological Imagination*. New York: Oxford University Press.

Moore, J. (2015). *Capitalism in the Web of Life: Ecology and the Accumulation of Capital*. New York: Verso Books.

Rafter, N. (2006). *Shots in the Mirror: Crime Films and Society*. 2nd ed. New York: Oxford University.

Shone, T. (2004). *Blockbuster: How Hollywood Learned to Stop Worrying and Love the Summer*. London: Simon & Schuster.

Spivak, G.C. (1999). *A Critique of Postcolonial Reason: Toward a History of the Vanishing Present*. Cambridge, MA: Harvard University Press.

Spivak, G.C. (2012). *An Aesthetic Education in the Era of Globalization*. Cambridge, MA: Harvard University Press.

Vogler, C. (1992). *The Writer's Journey: Mythic Structure for Storytellers and Screenwriters*. Studio City, CA: Michael Wiese Productions.

Part II

The Case

4 Story

Alfonso Cuarón's *Children of Men* (2006) is an adaptation of *The Children of Men*, a novel by P.D. James (1992), who was best known for her Adam Dalgliesh crime fiction series. After a premiere at the 63rd Venice International Film Festival, the film was released in the United Kingdom in September and in the United States in December. The initial release in the United States was limited to 16 theatres but expanded to over 1,000 in January 2007 (IMDbPro 2025). *Children of Men* was widely reviewed; lauded by influential critics such as Roger Ebert, Manohla Dargis, Peter Travers, Dana Stevens, and Kenneth Turan; and scored well on platforms such as *Rotten Tomatoes* and *Metacritic*. Notwithstanding, the film was a commercial failure, grossing only $70 million of the $76 million it cost to make. There was something, it seemed, that failed to resonate with audiences, and its poor performance at the box office was mirrored by a disappointing performance at the Academy Awards and British Academy Film (BAFTA) Awards. Although nominated for three of each, *Children of Men* received no Academy Awards and only two BAFTAs, for Best Cinematography and Best Production Design (Jacobson 2020). *Children of Men* was released on DVD in the United Kingdom in January and in the United States in March, but sales were unremarkable and by April, it seemed doomed to obscurity.

4.1 Content

Cinema is a hybrid mode of representation that combines the pictorial, visual, or depictive with the linguistic, verbal, or descriptive and the only difference between feature films and television series within the cinematic mode of representation in the 21st century is the way in which the two types of representation are typically consumed (i.e. in a single sitting as opposed to multiple viewings at intervals). Across all modes of representation, the form of a particular representation can be distinguished from the content of that representation and the significance of the relationship, which is also referred to as style and substance, manner and matter, and medium and message, has been recognised for over two millennia. Broadly but accurately speaking, form is *how* a representation represents and content is *what* a representation represents. The

DOI: 10.4324/9781003565826-6

integration of form and content, of creative and inventive design with a manifold and overdetermined construct, has been identified as both valuable in itself (often referred to as the artistic or aesthetic value of a work or text) and valuable as a means to an end (often ethically or politically, as in aesthetic education). The final and instrumental values of the relation are based on the way in which form and content combine to reveal the universal – a concrete or abstract object or image – in the particular (Plato 1997; Schiller 1795; Bradley 1909; McGregor 2016).[1] While form is specific to each mode of representation, content remains constant across all modes and includes subject, theme, character, setting, and action.

Children of Men is 100 minutes from opening to closing credits and has the five-act structure characteristic of Hollywood blockbusters: exposition, rising action, climax, crisis, and resolution. The length of the acts reflects the five-act structure as an extension of the older, traditional three-act structure of exposition, complication, and resolution, in which the complication is itself complicated to accommodate changes of both direction and pace. The first, second, and fifth acts are approximately 15 minutes long and the third and fourth approximately 25 minutes long, creating a complication of nearly 70 minutes in length (McGregor 2021; Grčki & McGregor 2024). The narrative begins in London on 16 November 2027, 21 years in the future at the time of the film's release in 2006, during an extinction event caused by the inability of human beings to reproduce. The global response to the infertility pandemic, which followed an influenza pandemic in 2008, has been economic and political collapse. The subsequent global polycrisis has only been averted in the United Kingdom by a suspension of democracy and the establishment of a police state, placing the film firmly in the dystopian science fiction genre. The opening credits conclude with a news broadcast announcing the murder of Diego Ricardo, the youngest living person, in Buenos Aires. Ricardo was the last human being to be born, in 2009, and 18 at the time of his death. The credits segue to the opening scene in which a crowd of people are watching the news in a coffee shop. After a few seconds, Theo Faron (played by Clive Owen) enters the shop and pushes his way to the counter, his lack of both interest and emotion juxtaposed with those of the rest of the customers. Theo leaves with his coffee, stops to top it up with alcohol, and is replacing the lid on the cup when a bomb detonates in either the coffee shop or the shop next door (it is not clear which).

In less than two and a half minutes (less than one and a half excluding the credits), the audience has been presented with almost all they need to know about two crucial elements of the narrative: the setting and the character of the protagonist. The rest of the world, including the United States, is in a state of chaos and conflict; the United Kingdom closed its borders in 2019, maintains a military occupation of its mosques, and is host to at least one insurgency (whether Islamist or not is, like the precise location of the bomb, unclear). Theo is a functioning alcoholic with little or no hope for the future of humanity, disinterested and emotionally numb. In the next scene, he is revealed as a civil servant at the Ministry of Energy who takes no pleasure in his work and has no close relationships with his colleagues.[2] Theo asks his manager if he can take compassionate leave following Ricardo's murder and the brief conversation very quickly confirms the audience's initial impression of him: he is unmoved by the murder and by his close proximity to the bomb blast and happy to exploit the international outpouring of grief to make his own life more tolerable. In the final scene of the film, Theo rows Kee (played by Clare-Hope Ashitey) and her baby daughter, the first child to be born in 18 years, to safety, concluding a lengthy and costly escape from both the authorities and enemies of the state. In a classic twist in the tail of the tale, Theo admits to Kee that he has been mortally wounded and is dead by the final shot of the film, which is of a fishing boat arriving to rescue Kee and her daughter. As such, the plot of *Children of Men* appears to move from an inaugural condition in which a cynical, dispassionate Theo lives a meaningless life to a retrospectively inevitable condition in which Theo makes his life meaningful by sacrificing it for the future of humanity.

The exposition (0 to 16 minutes) introduces Theo, Jasper Palmer (played by Michael Caine), Julian Taylor (played by Julian Moore), and Luke (played by Chiwetel Ejiofor). Theo is a former political activist and was married to Julian, with whom he had a child. The couple separated after Dylan's death from influenza in 2008, and Julian has remained true to their ideals, assuming leadership of an allegedly non-violent group of pro-democracy activists called the Fishes, with Luke as her deputy. Jasper is Theo's only friend, an ageing hippy who lives in isolation in the countryside with a severely disabled wife, Janice (played by Philippa Urquhart), and survives by growing and selling cannabis. The plot is set in motion when Julian asks Theo to use his family connections to acquire transit papers that will enable an illegal immigrant (all immigration has been criminalised since 2018) to reach the coast and escape the United Kingdom before she is arrested and interned in a concentration camp. In the first instance of what will become a recurring feature of the film, a significant event – Theo's decision to assist Julian – occurs offscreen.

The rising action (16 to 30 minutes) begins with Theo on route to meet his cousin, Nigel (played by Danny Huston), who is the Minister of Art and director of the Ark of the Arts, which was established to protect the world's greatest artworks from destruction as global civilisation collapsed. Once again offscreen, Theo secures transit papers for two people, which means that he will have to travel with the immigrant and that he can demand more money for his services to the Fishes. Julian introduces Theo to Kee, and the three of them begin their journey, accompanied by Miriam (played by Pam Ferris), a former midwife, and driven by Luke. En route to Canterbury, they are ambushed by a large gang, who shoot Julian. Their escape is nearly foiled by a pair of police officers, but Luke kills them.

The climax (30 to 59 minutes) begins with Julian's death and is followed by the arrival of Theo, Kee, Luke, and Miriam at a farm the Fishes are using as a safehouse. Luke is appointed to replace Julian and Theo makes a series of discoveries: first, that Kee is pregnant, then that Julian's plan was to send her to the Human Project, a nonstate scientific organisation rumoured to be researching a cure for infertility, and finally that Luke orchestrated Julian's death so that he could assume the command of the Fishes and use Kee's child to negotiate with the government. Theo, Kee, and Miriam flee from the Fishes and take refuge with Jasper. Jasper exploits his relationship with Syd (played by Peter Mullan), a corrupt officer in the paramilitary Homeland Security agency, to arrange their escape through Bexhill-on-Sea, which has been transformed into a concentration camp. When the Fishes discover Jasper's hideaway, he tells Theo, Kee, and Miriam to leave, euthanises Janice, and confronts Luke on his own. Theo watches Luke kill Jasper from a vantage point and the act ends with his third escape.

The stakes are heightened even further in the crisis (59 to 85 minutes), when Kee goes into labour as she, Theo, and Miriam approach Bexhill on a Homeland Security bus. Miriam distracts the Homeland Security officers from noticing Kee's condition by feigning a mental health emergency and is separated from the rest of the immigrants for what appears to be either solitary confinement or execution. Theo and Kee are processed into the camp, which is run entirely by the immigrants themselves, many of whom belong to rival factions. Theo assists Kee to give birth to a baby girl she subsequently names Dylan. The following day, the Fishes blow a hole in the camp wall and initiate a multi-sided armed conflict that is aggravated by the deployment of the military. Luke captures Kee and Dylan and orders the execution of Theo.

The resolution (85 to 100 minutes) begins with Theo's escape and presents him with three challenges, each of which is potentially more difficult than its predecessor. First, he must rescue Kee from Luke, then reach the boat that awaits him and Kee, and finally rendezvous with the Human Project vessel. Luke is killed in combat and the Army allow Theo and Kee to leave when they see she is carrying Dylan. As Theo rows the boat out to a buoy, the Air Force launches a strike on Bexhill, and Kee learns that he is mortally wounded. Moments after Theo dies, Kee sees the Human Project boat and the film ends *in media res*, before she is picked up.

4.2 Form

As noted above, form, unlike content, is specific to each mode of representation. In hybrid modes of representation, such as cinema, form consists of elements from the constituent modes of representation as well as those unique to the hybrid mode itself. The cinematic mode of representation is, as also noted above, a hybrid of both the pictorial, visual, or depictive and the linguistic, verbal, or descriptive. In the Western tradition of the fine arts, cinema typically combines painting (the screen) with poetry (the script), dance (the acting), and music (the score). The formal elements of these discrete categories include (Pater 1893; Gombrich 1950; Langer 1953; McGregor 2016): line, shape, form,[3] colour, and texture (painting); structure, morphology, syntax, metre, and tropes (poetry); structure, energy, space, time, and clarity (dance); and sound, harmony, melody, rhythm, and form (music). In consequence of the multiplicity of forms associated with the cinematic mode of representation, the convention is to separate film form from film style, where the former refers to the formal elements associated with films as a narrative and the latter to the formal elements associated with the technology required to animate and articulate pictures. Film form includes genre, structure, function, and framework. Film style includes *mise-en-scène*, cinematography, editing, and sound. This conventional distinction is a convenient way to approach what is probably the most complex of all modes of representation, but the omission of a key component of cinema in acting is worth bearing in mind (Thompson 1988; McGregor 2021; Bordwell, Thompson & Smith 2023). Although 'narrative representation' is often employed as if it is a mode of representation, 'narrative' is more accurately understood as a formal element, as narrative form is imposed on a substantive sequence of events. Narrative form in this sense is often referred to as *plot* and *emplotment* is the process of representing a sequence of events as a narrative or, in more colloquial terms, a story.[4] While I shall discuss elements of both film form and film style in all three chapters of this part of the essay, this section will deal primarily with the narrative form of *Children of Men*.

With respect to fiction, 'genre' can denote any of a category, kind, form, style, or purpose of a text or work. Within the linguistic, verbal, or descriptive mode of representation, it is often also used as a catch-all term for particular texts or works that either do not aspire to or aspire but fail to achieve the status of art, that is genre fiction versus literature (which includes poems, plays, and novels). In criminology, there is a similar distinction between art and popular culture (although 'popular culture' includes much else in addition to fiction). What is significant about genre for my purposes is that each genre is constituted by a set of conventions and those conventions create a set of expectations in audiences. In films aimed at a mass market audience, those expectations are almost always met and, increasingly in the 21st century, produce sequels or franchises when they are commercially successful. Art films, which are typically aimed at smaller, more sophisticated audiences, are usually created in isolation from genre conventions, although directors may deploy them for the purpose of either exploiting or subverting audience expectations (Brisman & South, 2014, 2025; Brisman 2016; McGregor 2016, 2021). I identified *Children of Men* as a dystopian science fiction in virtue of its setting, but the synopsis provided above shows that it can be more accurately described in generic terms as a dystopian thriller, its narrative consisting of a series of escapes: from the ambush, from the farm, from Jasper's house, and from Bexhill Refugee Camp.

The intimacy of the relation between the form and the content of *Children of Men* was demonstrated in my synopsis as I could not help but structure the film's content in some way. At the most basic level, the narrative consists of five acts: exposition, rising action, climax, crisis, and resolution. *Children of Men* has a particularly complex structure, however, because two superstructures are built on the basic five-act structure and a stylistic feature is accorded unusual structural significance. The first of the two superstructures is the monomyth, which I discussed in Chapters 2 and 3, and to which I return below. The second is allegory, which I also discussed in Chapters 2 and 3 and to which I return in Chapter 6. The stylistic feature is the film's cinematography – the work of Emmanuel Lubezki, a frequent collaborator with both Cuarón and Terrence Malick – specifically two long takes or single-shot sequences, one at the end of the second act and the other providing the segue from the fourth to the final act.[5] One does not need to be a structuralist the likes of Ferdinand de Saussure (1916), Claude Lévi-Strauss (1949), Roland Barthes (1957), or Jacques Lacan (1975) to be able to recognise that structure, to at least some extent, determines meaning and, indeed, one of the most compelling applications of the form–content relationship to poetry is called 'resonant meaning,' where resonance is form and meaning content (Bradley 1909: 14). The same is true of the narrative location of – or, more accurately, the juxtaposition between – the two long takes and their significance extends beyond their relationship to one another to their relationships with the five-act, monomythic, and allegorical structures of the film. I return to the long takes in Chapter 5.

The idea that everything presented in a narrative representation is presented for a purpose and thus has a function (or, more often, multiple functions) is integral to the concept of narrative. I described narrative as a form that is imposed on a sequence of events. A historical narrative, for example, is not simply a list or even a chronicle of what happens but a selection of characters, settings, and actions that have relevance or consequence to what the narrative is about, that is its patterns of meaning or thematic perspective. Similarly, *Children of Men* does not present *everything* Theo does in the seven days that begin with him buying a cup of coffee and end with his death, only those relevant and of consequence to the story Cuarón is telling.[6] The relationship between selectivity and function is crucial to narrative, which can be described as following the principle of functionality: every element of the text or work, including all those associated with form, style, acting, and content, has a function in the representation as a unified whole. In the case of a feature film, the creation of which requires hundreds if not thousands of individual contributions, control of functionality will usually be by the director. To take a banal yet significant example, one of the reasons that the middle of the first act features Theo's visit to Jasper is to introduce him and his relationship with Theo before Theo arrives with Kee and Miriam in the third act. The climax or turning point of the narrative (which I discuss in the next section) takes place at Jasper's house and would require more exposition – which would, in turn, slow its pace – were audiences not already familiar with Jasper. The scene and Jasper serve many other functions, but that is perhaps the most basic (Auster 1987; Lamarque 2009; McGregor 2016; Rabiger & Hurbis-Cherrier 2020).

Finally, narrative representation is necessarily rather than contingently perspectival – or, more specifically, complex narrative representations such as films, plays, and novels are essentially perspectival.[7] When narrative form is imposed on a sequence of events, it not only selects some but not all of those events but also represents them – as well as the people and places represented – from a particular perspective. Narratives may of course present people, places, and events from multiple points of view, which may be more obvious in the cinematic mode of representation than others because audiences are often aware of the movement of the camera (or at least the apparent movement of the camera). These points of view or perspectives internal to the narrative combine to create the external (or, in the case of a feature film, directorial) perspective that constitutes the narrative itself. To take another banal yet significant example from *Children of Men*, London in 2027 is represented as a significantly worse place to live than London in 2006, in spite of many similarities between them, including the threat of insurgent attacks. This is, of course, because the film is a dystopian thriller, set in a future that is worse rather than better than the present. The (external) perspective constituting a narrative representation produces a narrative framework, which invites audiences to adopt a certain set of emotional responses and evaluative attitudes. The framework foregrounds the essentially ethical character of emplotment as narrative movement from an 'unstable inaugural condition, a condition that *is* but *ought not* . . . through a process of sifting and exploration in search of an unknown but retrospectively inevitable condition that *is* and truly *ought-to-be*' (Harpham 1999: 36, emphasis in original; see also Currie 2010; Smith 2022; McGregor 2021). The narrative movement of *Children of Men* is thus, as noted above, from an inaugural condition in which a cynical, dispassionate Theo leads a meaningless life (*is-but-ought-not-to-be*) to his decision to dedicate himself to Kee's protection when he discovers her pregnancy to a retrospectively inevitable condition in which Theo makes his life meaningful by sacrificing it for the future of humanity (*is-and-ought-to-be*).

4.3 Meaning

The perspective of a narrative constitutes the representation and produces the framework that, in turn, shapes not only the emotions and attitudes of the audience, but also their desires. Unless we reject the framework of *Children of Men* (which we might do for several reasons), we experience sadness when the narrative closes because we approve of Theo's progress from nostalgic hedonism to self-sacrifice for the future of humanity and because we want him to succeed in his mission – to both save Kee and Dylan *and* escape with them. The audience's emotions, attitudes, and desires are inextricably bound to the various levels of meaning of and in the narrative, where the integration of a creative and inventive structure with a manifold and overdetermined subject develops patterns of thematic meaning that reveal the universal in the particular. The relationship between structure and form on the one hand and meaning and theme on the other is extremely complex and the most obvious place to begin is with the narrative's monomythic structure. Applied to *Children of Men*, Joseph Campbell's (2008) summary of the monomyth might read:

> Theo ventures forth from the world of physical and spiritual infertility into a region of revolutionary hope and commitment: Luke and the Fishes are there encountered and a decisive victory is won when Theo helps Kee escape their clutches: Theo comes back to humanity with the power to save Kee and her newborn child.

Theo's coming back is psychological rather than physical, but his hero's journey is nonetheless consistent with Campbell's model.

In Chapter 2, I set out John Yorke's (2013; see also Grčki & McGregor 2024) five-point template for monomythic structure as: call to action, initial objective achieved, the point of no return, all is lost (or won), and resolution in victory (or defeat). The call to action takes place in the last third of the first act, when Theo is kidnapped by the Fishes and asked to secure transit papers by Julian (14:40). He initially refuses, and his answer to the call – the decision to ask Nigel for the papers – takes place offscreen, in between the first and second acts. The initial objective is achieved at the end of the second act, when Theo, Kee, Luke, and Miriam escape both the gang and the police (30:22). The turning point occurs at the midpoint of the narrative, two-thirds of the way into the third act, when Kee initiates her first casual conversation with Theo (50:38). She has previously regarded him with hostility and suspicion and the conversation that follows marks the beginning of their friendship. Recall from Chapter 2 that the turning point is also the point of no return when the narrative transforms from being driven by what the protagonist wants to what the protagonist needs. What Theo wants is to live what remains of his life in a state of numb neutrality, without emotional or moral commitment and the pain they cause and seeking solace in the memories of his life with Julian before the pandemic. He is called from this life by Julian and his answering of that call is because of his nostalgia for and desire to replicate his past. What Theo needs is to regain his humanity, allow himself to feel again, and commit to a cause that will make the rest of his life meaningful. All is lost at the end of the fourth act, when Luke captures Kee and Dylan and orders his lieutenant, Patric (played by Charlie Hunnam), to execute Theo (85:40).

The resolution is in victory, in the last 30 seconds of the film and after Theo's death. Kee, who is – along with Dylan – uninjured by her ordeal, sees the Human Project boat (which is offscreen at the time) – and instinctively reaches for Theo's corpse (99:44).

This monomythic interpretation suggests the significance of the themes of hope, commitment, and spirituality and the significance of these themes can be evaluated more closely by comparing *Children of Men* to the novel on which it is very loosely based, James' (1992) *The Children of Men*. Not without justification, Cuarón (cited in Riesman 2022) initially dismissed the novel with: 'I was not interested in a science-fiction thing about upper classes in a fascist country.' The novel was published in 1992 and set 19 years in the future, in 2021. Although the setting is broadly similar, the novel is driven by the relationships among Theo, Julian (who is pregnant), and Xan Lyppiatt (Theo's cousin and the Warden of England) rather than between Theo, Kee, and Luke. In the film, Julian's function is largely limited to the recruitment of and motivation for Theo and Theo's cousin, Nigel, is a minor character with Lyppiatt's role as primary antagonist being assumed by Luke. There is no refugee crisis in the novel, where immigrants are encouraged to maintain the English workforce and forcibly repatriated when they turn 60. Cuarón's (cited in Hennerson 2006) comment on the significance of Kee as not only a refugee, but a Black refugee is enlightening, particularly when read alongside his criticism of James: 'We're putting the future of humanity in the hands of the dispossessed and creating a new humanity to spring out of that.'

'P.D. James' is the penname of Phyllis Dorothy James (1920–2014), who was best known for her Adam Dalgleish mysteries, 14 novels published from 1962 to 2008. She published five other novels, all of which were also crime fictions except for *The Children of Men*, which is usually described as dystopian satire. Dalgleish is a high-ranking police officer, published poet, drives a Jaguar, and is indicative of James' own social circle. Her full title is in fact The Right Honourable The Baroness James of Holland Park OBE FRSA FRSL as she was created a life peer in 1991 (two years after publication of the eighth Dalgliesh novel and the year before publication of *The Children of Men*), sat in the House of Lords as a representative of the Conservative Party, and was a multimillionaire at the time of her death. James was deeply conservative politically and her conservatism extended to religion, where she was a devout member of the Church of England and a lay patron of the Prayer Book Society, an Anglican charity (*The Telegraph* 2014, 2015; *Church Times* 2014). Given both this background and the title of the novel, which is from *The Book of Psalms* in *The Holy Bible*, it is not surprising that it is peppered with Christian imagery and an explicit – even didactic – exploration of the themes of faith, hope, and redemption. The closing paragraph provides one of many examples. Julian has just given birth (to a boy, another important difference from the film) and asks Theo (who is not the father) to christen the child (James 1992: 342):

> The towel between her legs was heavily stained. He removed it without revulsion, almost without thought and, folding another, put it in place. There was very little water left in the bottle, but he hardly needed it. His tears were falling now over the child's forehead. From some far childhood memory he recalled the rite. The water had to flow, there were words which had to be said. It was with a thumb wet with his own tears and stained with her blood that he made on the child's forehead the sign of the cross.[8]

In Cuarón, the Christian themes are not only much more subtle but also diluted, speaking to a multitude of faiths or even a more general spirituality rather than any one religion. In his commentary on the film, Slavoj Žižek (cited in *Children of Men* 2007, 04:48–05:07, emphasis in original) praises it for precisely this subtlety, which is part of what Gregory Wolmart (2017) refers to as his anamorphic reading (which I discuss in Chapter 5):

> The fertility is spiritual fertility, it's to find the meaning of life and so on – so these are the reasons, again, for my admiration of the film, that precisely because it doesn't *directly* make some kind of political, moralistic parable and so on and so on, it works, perfectly.

Žižek (2008: 28) reflects on this spiritual infertility in his long essay *Violence: Six Sideways Reflections*, where he describes it in terms of Friedrich Nietzsche's (1883) conception of the *last man*, 'an apathetic creature with no great passion or commitment.' What is particularly interesting here (and a point to which I also return in Chapter 5) is that the full title of Francis Fukuyama's (1992) *magnum opus* in praise of late capitalism, often abbreviated to *The End of History*, is in fact *The End of History and the Last Man*. Although I disagree with Fukuyama's theses on both late capitalism and the film, the latter is particularly interesting in reading *Children of Men* as an allegory. Žižek is one of the seven experts interviewed in *The Possibility of Hope* (2007), a 27-minute documentary directed by Cuarón and included (along with Žižek's separate commentary) in the DVD releases of *Children of Men*. The documentary is divided into five chapters (Reality, Fear, Walls, Fever, and Hope); mixes shots from the film with stock footage; and includes commentary on the multiple crises of late capitalism, on *Children of Men*, and on the relationship between the two. Žižek (cited in *Children of Men* 2007, 05:32–05:38) is both the most optimistic about the future and the most lavish with his praise: 'Only films like this can guarantee that cinema as art will really survive.'

Cuarón's (cited in Guerrasio 2006) own comments on what appears to be the most important theme in the film are more ambiguous:

> We wanted the end to be a glimpse of a possibility of hope, for the audience to invest their own sense of hope into that ending. So if you're a hopeful person you'll see a lot of hope, and if you're a bleak person you'll see a complete hopelessness at the end.

This is, I think, disingenuous. While there are always audiences that will, for a wide variety of reasons, reject a narrative framework, *Children of Men's* perspective on hope is transparent (while remaining subtle) rather than opaque, as the attention to the sound effects of the final few seconds reveals. While a particularly uplifting part of the film's score plays in the background, Kee sings Dylan a lullaby, the bell on the buoy rings, and the waves lap gently (Tavener 2006). The imagery is neither as overt nor as obvious as the conclusion of the novel, but it clearly represents hope, spirituality, and renewal rather than their opposites. The final shot is replaced by a black screen, and there are sounds of children laughing and playing which segue to and through the credits. A second later, 'CHILDREN OF MEN' appears in white text for nine seconds before the credits roll (*Children of Men* 2006, 100:21–100:29). Any interpretation of this as providing a despairing perspective on the sequence of events is simply implausible. The film's theme is the triumph of hope over despair and the hero's journey has ended in physical (rescuing Kee and Dylan) and psychological (commitment to the future) victory, albeit one that has cost Theo his life.

Notes

1 The relationship between universals and particulars is one way of articulating the relationship between reality and its representation, which is discussed across the next three chapters.

2 The levels of narrative analysis I use are, from the specific to the general, shot, sequence, scene, and act. This is of course contrary to the shot, scene, and sequence used by filmmakers in creating and producing cinematic narratives, but I prefer the former for analysis and evaluation as it foregrounds the affinity with other narrative fictions.

3 'Form' can be ambiguous, especially when one is discussing multiple modes of representation. In addition to my delineation of form as how a representation represents, it can refer to: a category of art (such as painting, poetry, dance, and music); a three-dimensional shape (in painting); the structure of a composition (in music); and the elements of narrative (in film).

4 I shall use the two terms as synonyms, the significant distinction being between a sequence of events or series of episodes, incidents, or occurrences on the one hand and a narrative or story on the other.

5 There are at least four long takes in the film that have received critical attention, but these are the two longest.
6 The duration of the sequence of events of *Children of Men* is not explicit but is somewhere in the range of five to nine days. I have selected seven as the most likely.
7 For the distinction between complex, exemplary, or bounded narrative representations on the one hand and basic, simple, or notional narrative representations on the other, see Presser (2018) and McGregor (2021).
8 Notwithstanding the significant religious and political differences between James and Cuarón, she appears to have approved of his adaptation, with a prominent albeit uncredited appearance as a coffee shop customer (holding a dog) in the opening shot of the film (IMDb 2025).

References

Auster, P. (1987). *The New York Trilogy*. London: Faber & Faber.

Barthes, R. (1957/2000). *Mythologies*. Trans. A. Lavers. London: Vintage Books.

Bordwell, D., Thompson, K. & Smith, J. (2023). *Film-Art: An Introduction*. 13th ed. New York: McGraw-Hill Education.

Bradley, A.C. (1909/1959). *Oxford Lectures on Poetry*. London: Macmillan Publishing.

Brisman, A. (2016). On Narrative and Green Cultural Criminology. *International Journal for Crime, Justice and Social Democracy*, 6 (2), 64–77.

Brisman, A. & South, N. (2014). *Green Cultural Criminology: Constructions of Environmental Harm, Consumerism, and Resistance to Ecocide*. Abingdon: Routledge.

Brisman, A. & South, N. (2025). *Monstrous Nature and Representations of Environmental Harms: A Green Cultural Criminological Perspective*. Philadelphia, PA: Temple University Press.

Campbell, J. (2008). *The Hero with a Thousand Faces*. 3rd ed. Novato, CA: New World Library.

Children of Men (2006). Directed by Alfonso Cuarón. US: Universal Pictures.

Children of Men – Comments by Slavoj Žižek (2007). In: *Children of Men – Special Edition*. Directed by Alfonso Cuarón. DVD, B000NIMZZM. Universal City, CA: Universal.

Church Times (2014). Obituary: Baroness James of Holland Park. 5 December. Available at: <www.churchtimes.co.uk/articles/2014/5-december/gazette/obituaries/obituary-baroness-james-of-holland-park>.

Currie, G. (2010). *Narratives and Narrators: A Philosophy of Stories*. Oxford: Oxford University Press.

Fisher, M. (2022). *Capitalist Realism: Is There No Alternative?* 2nd ed. Alresford: Zero Books.

Fukuyama, F. (1992). *The End of History and the Last Man*. London: Penguin Books.

Fukuyama, F. (2016). My Favorite Movie with Francis Fukuyama: Children of Men. *Future Tense*. Washington, DC: New America. Available at: <www.youtube.com/watch?v=03SrMeLmOUc>. Accessed 31.10.24.

Gombrich, E.H. (1950). *The Story of Art*. London: Phaidon Press.
Grčki, D. & McGregor, R. (2024). *An Epistemology of Criminological Cinema*. Abingdon: Routledge.
Guerrasio, J. (2006). A New Humanity. 22 December. *Filmmaker*. Available at: <https://web.archive.org/web/20081009182846/www.filmmakermagazine.com/archives/online_features/hopeless_future.php>.
Harpham, G.G. (1999). *Shadows of Ethics: Criticism and the Just Society*. Durham, NC: Duke University Press.
Hennerson, E. (2006). Brave New World. 19 December. *Los Angeles Daily News*. Available at: <https://web.archive.org/web/20070930201349/www.dailynews.com/entertainment/ci_4866854>.
IMDb (2025). *Children of Men*. Available at: <www.imdb.com/title/tt0206634/>.
IMDbPro (2025). Children of Men. *Box Office Mojo*. Available at: <www.boxofficemojo.com/release/rl1464567297/>.
Jacobson, G. (2020). Why Children of Men Haunts the Present Moment. 22 July. *The New Statesman*. Available at: <www.newstatesman.com/culture/film/2020/07/children-of-men-alfonso-cuaron-2006-apocalypse-coronavirus>.
James, P.D. (1992/2018). *The Children of Men*. London: Faber & Faber.
Lacan, J. (1975/1998). *The Seminar of Jacques Lacan Book I: Freud's Papers on Technique 1953–1954*. Trans. J. Forrester. New York: W.W. Norton & Company.
Lamarque, P. (2009). *The Philosophy of Literature*. Malden, MA: Blackwell Publishing.
Langer, S.K. (1953). *Feeling and Form: A Theory of Art*. New York: Charles Scribner's Sons.
Lévi-Strauss, C. (1949/1969). *The Elementary Structures of Kinship*. Trans. anonymous. Boston, MA: Beacon Press.
McGregor, R. (2016). *The Value of Literature*. London: Rowman & Littlefield International.
McGregor, R. (2021). *A Criminology of Narrative Fiction*. Bristol: Bristol University Press.
Nietzsche, F. (1883/2006). *Thus Spoke Zarathustra: A Book for All and None*. Trans. A. del Caro. Cambridge: Cambridge University Press.
Pater, W. (1893/1980). *The Renaissance: Studies in Art and Poetry*. Berkeley, CA: University of California Press.
Plato (1997). Republic. Trans. G.M.A. Grube & C.D.C. Reeve. In: Cooper, J.M. (ed.). *Plato: Complete Works*. Indianapolis: Hackett Publishing, 971–1223.
The Possibility of Hope (2007). In: *Children of Men – Special Edition*. Directed by Alfonso Cuarón. DVD, B000NIMZZM. Universal City, CA: Universal.
Presser, L. (2018). *Inside Story: How Narratives Drive Mass Harm*. Oakland, CA: University of California Press.
Rabiger, M. & Hurbis-Cherrier, M. (2020). *Directing: Film Techniques and Aesthetics*. 6th ed. Abingdon: Routledge.

Riesman, A. (2022). Why 'Children of Men' Remains Relevant with Each Passing Year. 18 November. *Special Broadcasting Service*. Available at: <www.sbs.com.au/whats-on/article/why-children-of-men-remains-relevant-with-each-passing-year/n3tnyumqr>.

Saussure, F.de (1916/2020). *Course in General Linguistics*. Trans. R. Harris. London: Bloomsbury Publishing.

Schiller, J.C.F. von (1795/1967). *On the Aesthetic Education of Man: In a Series of Letters*. Trans. E.M. Wilkinson & L.A. Willoughby. London: Oxford University Press.

Smith, M. (2022). *Engaging Characters: Fiction, Emotion, and the Cinema*. 2nd ed. Oxford: Oxford University Press.

Tavener, J. (2006). *Fragments of a Prayer*. 12 December. US: Varèse Sarabande.

The Telegraph (2014). PD James – Obituary. 27 November. Available at: <www.telegraph.co.uk/news/obituaries/11258055/PD-James-obituary.html>.

The Telegraph (2015). Author PD James amassed £22 Million Fortune before She Died. 6 September. Available at: <www.telegraph.co.uk/culture/books/booknews/11847519/Author-PD-James-amassed-22-million-fortune-before-she-died.html>.

Thompson, K. (1988). *Breaking the Glass Armor: Neoformalist Film Analysis*. Princeton, NJ: Princeton University Press.

Wolmart, G. (2017). On Anamorphic Adaptations and the Children of Men. *International Journal of Žižek Studies*, 11 (2), 1–21.

Yorke, J. (2013). *Into the Woods: How Stories Work and Why We Tell Them*. London: Penguin Books.

Žižek, S. (2008). *Violence: Six Sideways Reflections*. New York: Picador Books.

5 Style

As is so often the case, *Children of Men's* (2006) financial flop was not simply a function of its failure to engage or entertain audiences. The primary cause was a self-fulfilling loss of confidence in the film by Universal Pictures. The studio was unable to develop a coherent and compelling marketing strategy and had switched its considerable resources to Paul Greengrass' *United 93* by the time of *Children of Men's* United States release, which was on Christmas Day, an incongruous date for a film so gloomy, grim, and iconoclastic. *United 93* was both critically and commercially successful, grossing the budget of *Children of Men* while costing five times less to make (IMDbPro 2025). The problem for Universal was that *Children of Men* was science fiction without any futuristic trappings (if anything, technology has regressed), an action thriller without an action hero (Theo never uses a firearm and kills only one person), and a big budget production in which the most marketable star dies in the first half hour (Julianne Moore, who plays Julian). The trailers were particularly poor, featuring a voiceover from Owen and an upbeat soundtrack, and the posters even worse, featuring a golden foetus and the tagline 'He must protect our only hope' (Riesman 2022). Alfonso Cuarón had, furthermore, switched motherhood of that only hope from a well-educated, upper middle class, White woman in P.D. James' (1992) novel to an uneducated, undocumented, Black immigrant (Jacobson 2020). Neither the studio nor audiences were, it seems, ready for *Children of Men* in 2006 and 2007.

5.1 Capitalist Realism

The late Mark Fisher (1968–2017) is unique in the United Kingdom (and perhaps the Anglophone world) in being widely read and cited in spite of self-publishing his work and never holding a tenured academic position. Furthermore, he authored only three books, the most famous of which – *Capitalist Realism: Is There No Alternative?* (Fisher 2009, 2022) – is no more than a long essay at 81 pages. *Capitalist Realism* was followed by *Ghosts of My Life: Writings on Depression, Hauntology and Lost Futures* (Fisher 2014) and *The Weird and the Eerie* (Fisher 2016), both collections of short essays, the latter published

DOI: 10.4324/9781003565826-7

a month before his suicide. Fisher and novelist Tariq Goddard set up Zero Books as an imprint of John Hunt Publishing (now Collective Ink Limited) in 2007, following Fisher's unsuccessful approaches to publishers, and launched Repeater Books as an imprint of Watkins Media in 2014 (Goddard 2021). Prior to the publication of *Capitalist Realism* by Zero Books at the end of 2009, Fisher was best known for his blog, *k-punk*, which he started on 23 September 2003 and finished on 11 March 2013.[1] All three of his books drew heavily on his posts in *k-punk* and Repeater Books published the majority of the posts together with the unfinished manuscript of his next book, *Acid Communism*, as *k-punk: The Collected and Unpublished Writings of Mark Fisher (2004–2016)* (Fisher 2018b). Curiously, given that Fisher's (1998, 2018a) thesis was on cybernetic theory and horror fiction, he published an article in a criminology journal while a postgraduate student, 'White Magic' in *The Red Feather Journal of Postmodern Criminology: An International Journal*.[2]

Capitalist Realism is now in its second edition and remains Fisher's (2022) most popular and significant contribution to critical theory. The essay's argument is that late capitalism has created a situation in which it is impossible to even imagine a political economy beyond the market fundamentalism of neoliberalism. This is even more compelling now than at the time of publication, in consequence of the 24/7 access to social media provided by smartphones: social media achieved mainstream popularity with the launch of Facebook in 2004 and smartphones have become ubiquitous since 2011, outselling personal computers for the first time in the last quarter of 2010 (Arthur 2012; Ortiz-Ospina 2019). The content of *Capitalist Realism* can be divided into three parts. In the first (chapters 1 to 3), Fisher defines his title, delineates it as a concept, and identifies three aporias – inherent contradictions or paradoxes – within the concept. In the second and most substantial part (chapters 4 to 8), he draws on his experience as a teacher in further education to explore two of the aporias, the rise of mental health conditions and the rise of bureaucracy, before detailing the functioning of capitalist realism in first figurative and then literal terms.[3] Fisher concludes (chapter 9) by discussing prospects for an authentic education and a meaningful politics within the limits set by capitalist realism.

Fisher's critical method standardly involves beginning a chapter with a cinematic representation, carefully selected as exemplary of a particular aspect of capitalist realism, and employing it as a jumping-off point into – and lens for the analysis of – the philosophy, politics, or economics of late capitalism. The first chapter of *Capitalist Realism*, 'It's easier to imagine the end of the world than the end of capitalism', and thus the whole essay begins with *Children of Men*, which is used to introduce two issues. The first and less important refers to Theo's visit to Nigel at the Ministry of Art in Battersea Power Station. The ministry hosts the Ark of the Arts, whose collection includes Banksy's *Kissing Coppers* (2004), Michelangelo's *David* (1504), Picasso's *Guernica* (1937), and Algie, the inflatable pig on the cover of Pink Floyd's

Animals (1977) album. Fisher claims that the Ark dramatises the debate that inaugurated philosophical aesthetics as a field within philosophy, whether aesthetic properties exist in the object, in the eye of the beholder, or in both object and subject. By assigning all cultural artefacts a monetary value based on their market rather than cultural value, capitalism consumes both culture and history. Fisher (2022: 4) compares the British Museum with the Ark of the Arts: 'Walk around the British Museum, where you see objects torn from their lifeworlds and assembled as if on the deck of some Predator spacecraft, and you have a powerful image of this process at work.' Fisher's comments recall Slavoj Žižek (cited in *Children of Men* 2007, 02:23–02:49) on the despair characteristic of late capitalism:

> The true infertility is the very lack of meaningful historical experience and that's why I like this elegant point in the film of importing all the works of art. All those classical statues are there, but they are deprived of a world, they are totally meaningless because what does it mean to have a statue of Michelangelo or whatever?

Žižek uses *world* in the same sense as Fisher's *lifeworld*, which is perhaps best expressed by Martin Heidegger (1927) as neither the physical world in which human agency exists, nor the physical world as experienced by human agency, but a system of assignments of intersubjective meaning within which human agency projects itself and presses forward into possibilities. The cultural artefacts in both the Ark of the Arts and the British Museum have been severed from the system of assignments that underpin their production and reception and no longer have value as either works of art or as sites of intersubjective meaning.

The second and more important issue that Fisher (2022: 2, emphasis in original) raises through *Children of Men* is the definition of capitalist realism itself:

> Watching *Children of Men*, we are inevitably reminded of the phrase attributed to Fredric Jameson and Slavoj Žižek, that it is easier to imagine the end of the world than it is to imagine the end of capitalism.[4] That slogan captures precisely what I mean by 'capitalist realism': the widespread sense that not only is capitalism the only viable political and economic system, but also that it is now impossible even to *imagine* a coherent alternative to it.

Rather than an alternative to late capitalism, the dystopia in *Children of Men* is an extrapolation that examines what Jameson (1991: 46) refers to as 'the logic of late capitalism' in times of crisis. Capitalism is, as Fisher (2018b, 2022) points out, *always* in crisis and always *overcomes* crises by its apparently inexhaustible capacities for both adaptation to change and insinuation into every part of late modern life. He (Fisher 2022: 2) coins a nice clause of his own to describe both the world of *Children of Men* and the world in which we watch it: 'ultra-authoritarianism and Capital are by no means incompatible: internment camps and franchise coffee bars co-exist.'

Again, this recalls Žižek on the film, this time from his initial comments in *The Possibility of Hope* (2007), which open both the first chapter, 'Reality,' and the documentary.[5] Žižek draws on a point made in G.W.F. Hegel's philosophical aesthetics, that a good portrait looks more like its subject than the subject does herself and praises *Children of Men* for doing the same for the world in which it was produced and consumed. The idea is that there is some sort of essence or a crucial characteristic of the subject of a portrait that the artist first perceives, then represents, and finally communicates to those who view the painting, an essence or characteristic that one might fail to perceive by meeting the person themselves. In *Children of Men*, this quality of art is related to what Gregory Wolmart (2017) refers to as Žižek's anamorphic reading, which I discuss below. Fisher (2022: 1) similarly characterises the dystopia represented in the film, of co-existing franchise coffee bars and internment camps, as 'specific to late capitalism.' *Children of Men* thus exemplifies, dramatises, or stages capitalist realism, a situation in which capitalism has consumed the human imagination to the extent where the most one can hope for is to control the worst excesses of political economy rather than change it. Žižek (1999a) agrees with Francis Fukuyama (1992), describing the threat of an alternative to capitalism as disappearing in the early 1990s. Unlike Fukuyama, Žižek (1999a: 198, emphasis in original) is not optimistic about this foreclosure of the political, which he identifies as 'postmodern *post-politics.*'

As such, *Children of Men* is a site of synergy for not only Žižek and Fisher's critical theory but also that of Jameson (1981, 1991, 2002, 2005, 2007, 2013, 2019), in particular the theory of aesthetic education he sets out in *The Poetics of Social Forms*. Recall from Chapter 2 that aesthetic education is a misnomer because the concept concerns ethical and more importantly political education achieved by aesthetic or artistic means. The core thesis is that there is a relationship among the aesthetic, ethical, and political and the precise character of this relationship differs according to the Marxist, Hegelian, Aristotelian, and Levinasian strands. In Jameson's Marxist strand, novels provide new ways of thinking about the world in their form and aesthetic education is inseparable from capitalist realism, allegorical structure, and utopian desire. Recall from Chapter 4 that the form of cinematic narratives is particularly complex given the hybrid nature of the mode of representation, in consequence of which the counterpart claim would be that they provide new ways of thinking about the world in their form and in their style. I shall return to the former in the next chapter in which I also discuss what I believe is Fisher's (2022) most important claim about *Children of Men's* value, the way in which it represents not just the lived experience of catastrophe but also the lived experience of *living through* catastrophe. With respect to the relationship between new ways of thinking about the world and film style, I explore both *mise-en-scène* and cinematography in the remainder of this chapter.

5.2 Anamorphic Art

In stark contrast to Fisher, Žižek has published prolifically, for commercial as well as academic presses and in multiple languages, and held a multitude of prestigious academic positions since completing his first (of two) doctorates at the University of Ljubljana in 1981. In the 21st century – especially after the September 9/11 attacks – he became a global public intellectual and is, at the time of writing and in my opinion, the world's most famous living philosopher. Žižek's (1999b: ix) philosophy combines the works of two particularly complex theorists, Hegel and Jacques Lacan, and he provides a succinct summary of his overarching project as: 'the core of my entire work is the endeavour to use Lacan as a privileged intellectual tool to reactualize German Idealism.' Žižek's synthesis of German Idealism and psychoanalytic metapsychology draws on a vast number of complex sources in addition to Hegel and Lacan, including the *oeuvres* of Immanuel Kant, Friedrich Wilhelm Joseph Schelling, and Sigmund Freud (Johnston 2008; Bar-El 2023). One of the reasons for Žižek's popularity is his frequent use of examples from popular culture, especially films. Curiously, given his publication and institutional success, Žižek has had very little impact on criminology as a discipline and the most sustained engagement with his philosophy is in the ultra-realist theory of Steve Hall, Simon Winlow, and Thomas Raymen (Hall 2012; Hall & Winlow 2015; Raymen 2022), which deploys his (Žižek 1989, 1999a, 2006; Johnston 2008) transcendental materialist theory of subjectivity.

Wolmart (2017: 2) describes Žižek's reading of *Children of Men* as 'anamorphic' on the basis of Žižek's (cited in *Children of Men* 2007, 00:07–00:39, emphasis in original) commentary on the film:

> For me, *Children of Men* . . . I would say that the true focus of the film is there in the background and it's crucial to leave it *as* a background. Here comes his true art – Cuarón's – it's the paradox of what I would call an anamorphosis. If you look at the thing too directly, the oppressive social dimension, you don't see it. You can see it in an oblique way *only* if it remains in the background.

Although he does not use the term, Žižek (cited in *Children of Men* 2007: 01:40–02:04, emphasis in original) subsequently considers Cuarón's monomythic storytelling in terms of anamorphosis:

> It's not really that all this infertility and so on is just a pretext for . . . I don't know, the hero's inner journey from this apathetic antihero mode to more active engagement and so on – *no*, it's this fate of the individual

> hero remains a kind of a *prism* through which you see the background even more sharply.

The idea is related to Žižek's comment about realistic representation, of the portrait as providing something more than our own perception of its subject (in addition to its value in providing a likeness of people we have not met or will never have the opportunity to meet). Realism in representation, knowledge in art, and education by aesthetic means are often regarded as first functioning in an indirect way and then revealing a reality that is authentic rather than accurate and exceeds the purely perceptual capacities of those who experience it. With respect to *Children of Men*, Žižek's foregrounding of the background captures precisely what might be missed on an initial viewing of the film.

Cuarón's background is much more complex than his foreground, consisting of multiple layers that include plot, acting, sound, *mise-en-scène*, and cinematography. The plot, for example, is apparently simple: a single protagonist (Theo) and his allies make a series of attempts to escape from a single antagonist (Luke) and his allies over a week in a different but readily recognisable United Kingdom. This foreground, an uncomplicated instantiation of the monomyth, constitutes Žižek's prism, a superficial story that serves to sharpen an underlying plot, which I explore in Chapter 6. The complexity of the background might begin with the marketing problems with which I introduced this chapter. Theo is, at least apparently, the protagonist, but for the hero of an action thriller, he is remarkably passive: he remains unarmed the entire film, has no proficiency in or predilection for combat, and commits both of his acts of violence (killing Patric's cousin at 28 minutes and hitting Syd with what looks like a car battery at 80 minutes) in self-defence. Similarly, for a narrative set in the future, the level of technology has – as I noted above – regressed rather than progressed. Finally, both of the film's biggest stars, Julianne Moore and Michael Caine, are killed relatively early in the narrative (at 28 minutes and 59 minutes respectively). In Chapter 4, I noted the significance of the sound effects in the last sequence of the film and at the beginning of the end credits. A great deal of what is significant to not only the meaning of the narrative but even the plot takes place offscreen, in the background, or hidden in plain sight. The visual density of the background and the deployment of images to hide more than they reveal are both achieved by a combination of *mise-en-scène* and cinematography.

Children of Men was nominated for the Academy Award for Best Cinematography and the BAFTA Awards for Best Cinematography, Best Production Design, and Best Special Visual Effects and won the BAFTA for the first two. Emmanuel Lubezki's cinematography is so stunning as to almost defy description. Fisher (2018b: 179) and Gavin Jacobson (2020) both use the term *breathtaking*:

(1) Lubezki's cinematography is breathtaking. His photography seems to leech all organic and naturalistic vitality from the images, leaving them a washed-out grey-blue.

(2) Breathtaking cinematography is one reason for the film's enduring popularity. Rendered in washed-out palettes of grey, it was shot on handheld cameras, which lends the story propulsive energy and realism. The film is also famous for its single long takes.

I discuss Lubezki's two longest takes in the next section, but the relationship between the camera and the film set is no less remarkable in less dramatic shots and sequences. In his discussion of Universal Pictures' problems with marketing, Abraham Riesman (2022) mentions Cuarón's preference for a lack of exposition, for placing his audience *in medias res* without lengthy or tedious explanation. That may well be true, but much of the exposition of *Children of Men* is in the background. The background is, as one would expect, most often part of the *mise-en-scène*, for example the details of Theo's office, Jasper's house, and the room where Theo is interrogated by the Fishes in the first act. In a 2016 interview, Cuarón reported Lubezki as saying 'we cannot allow one single frame of this film to go without a comment on the state of things' (Martinez 2019; see also Riesman 2022). If one is watching the film on DVD or a streaming platform, stopping it at almost any point reveals a wealth of detail behind, around, and in front of the actors onscreen. The background is also, as I have already suggested, provided by the sound, which includes dialogue that seems unimportant but is actually expository. To take another example from the first act, the options for responsibility for the bomb blast (which reveal a great deal about the politics of 2027 United Kingdom) are very quickly set out by Jasper (a mere five minutes into the narrative) as: Islamists, the Fishes, or the government. Very often, subtle exposition like this is reinforced by subsequent *mise-en-scène*, sound, or both. Rudy Ralph Martinez (2019) highlights the significance of this when he states: 'Cuarón's usage of cross-reference is key.' To continue with the same example, Patric mentions the likelihood that the government planted the bomb nine minutes later, which recalls Žižek's (2008) concept of a tyrannical democracy that maintains its hegemony by maintaining a permanent state of emergency.

The conversations between Theo and Jasper and Theo and Patric are audible examples of what I described as hiding in plain sight. We hear the words, but the actors either place little emphasis on them or there are other images and sounds that demand our attention. One of the most striking examples of this is in the first long take, the sequence in which Theo, Julian, Luke, Kee, and Miriam are ambushed by a gang. As Luke reverses away from assailants on foot, a motorbike with two men on it, later revealed as Patric and his cousin, pursues the car. The cinematography, action, and acting all detract from a key point that is patently obvious on subsequent viewings: if the aim of the ambush is to kill or capture the occupants of the car, then the shooter, riding pillion on the motorbike, should aim for the driver of the car – that is, Luke – but Patric very carefully takes aim at Julian (28:21), who is in the front passenger seat, before he opens fire. His next target is Theo, which – again – makes no sense if the aim of the ambush is robbery, rape, or even murder. When Theo subsequently overhears Patric and Luke discussing the ambush, it comes as a surprise, but the targeting of Julian was visible from the beginning.

Kee's pregnancy is a similar albeit distinct example. When Theo meets Kee and she appears onscreen for the first time (25:36), she is sitting in the back of a car. She is also sitting the first time she appears outside the car (30:47), assisting Miriam with a funereal ceremony for Julian. The first time Theo and the audience see Kee full-length is when she asks to meet him in a barn. She appears in a middle-distance shot, wearing a shift and standing behind a group of calves that obscure the view of her pelvis and legs (35:26). Again, on a second viewing, it is quite obvious that Kee has a disproportionately large abdomen, but it is another minute before her pregnancy is revealed to first Theo and then the audience (36:36). A simple explanation is that Cuarón is, in both of my examples, using form and style to create an aesthetic experience for his audience by increasing the suspense and stakes in order to drive the plot forward. When Julian is killed, we wonder if it was a random act of violence or if the ambush was orchestrated by someone. When Theo meets Kee, we wonder why she is worth the risk the Fishes are taking as we have already seen dozens of immigrants in cages awaiting removal to concentration camps. The suspense does indeed work on this level and the answers to the original questions create new questions for the audience – why does Luke want to take charge of the Fishes and how will Theo help Kee escape now that she is bearing humanity's only hope for the future? The examples are also – and more significantly – part of Cuarón's use of the background as a means of sharpening the foreground, particularly with respect to the film's semantics, symbolism, and subversiveness, which I discuss in Chapter 6.

5.3 Long Takes

I have already mentioned *Children of Men's* single-shot sequences or long takes several times – it is difficult to avoid them given both their striking character and the extent to which they contribute to the aesthetic or artistic value of the film. There are at least four notable single-shot sequences, but I shall analyse the longest two, the ambush and the rescue. Long takes are one of the multiple means by which directors employ elements of film style to configure the experience of audiences. Like all stylistic devices, a long take can perform a limitless variety of functions, at least three of which Fisher (2018b: 180) mentions in his evaluation of their use in *Children of Men*:

> The meticulously choreographed long takes – technical feats of some magnitude – have justly been highly praised, and they are all the more remarkable because they go beyond the familiar role of simulating documentary realism to serve a political and artistic vision.

Long takes are frequently employed to shape an audience experience that is particularly intimate, intense, and impelling and *Children of Men* is no exception. A long take is, however, less about the actual camerawork than about what the audience sees and feels. In other words, a long take refers to a representation that looks like a single-shot sequence and is experienced as one by the audience as well as actual single-shot sequences. The ambush was in fact impossible to film in a single shot, but the seamless integration of the constitutive shots in the editing process created the illusion of a single shot with the result that both the ambush and the rescue are experienced as long takes and configure the audience's emotions, attitudes, and desires in a similar manner (Bielik 2006; Riesman 2022).

The first long take, which I have called the ambush, is widely agreed to be 247 seconds (4 minutes and 7 seconds) long (26:15–30.22). It occurs at the end of the second act of the film, with Theo and Kee en route to a checkpoint on the road to Canterbury. They are being escorted by Luke, Julian, and Miriam and the plan is that once the checkpoint has been reached, Theo and Kee will proceed to the coast on their own, using Theo's transit papers. The single-shot sequence opens with a close-up of Theo asleep in the backseat of the car as Luke drives along a rural road and most of the sequence is filmed from a point of view inside the car with close-ups of the occupants and with their views of what happens outside the car. The first minute and a half consists of an amusing and light-hearted conversation among all four passengers, dominated by Theo and Julian, who play an affectionate game in which they blow a table tennis ball into one another's mouths. Miriam then tells Luke to look out and the camera shows a burning car rolling downhill through the woods to block the road ahead. As Luke stops his car and puts it into reverse, dozens of people armed with sticks and rocks attack from both sides. Just under two minutes into the shot, a motorbike appears in front of Luke's car, which is still reversing, carrying two unidentified men (Patric and his cousin). Luke outdistances the gang, but the motorbike catches up and draws alongside. This is for Patric to see where everyone is sitting so that he can make sure of his target. A few seconds after the appearance of the motorbike, Patric fires a single shot that hits Julian in the throat. The motorbike then pulls alongside the car again, Patric takes aim at Theo, Theo opens the rear passenger door onto the motorbike, and it crashes. Two and a half minutes into the shot, Luke succeeds in turning the car around and Theo tries to stem the flow of blood from Julian's wound. Just before the three-minute mark, Julian appears to die, police sirens are heard, and two police cars and vans drive past, apparently in response to the ambush. One of the police cars turns around, pursues Luke's car, and orders him to pull over. Two police officers approach the car with weapons drawn, shouting instructions, while the occupants respond by shouting that they are British citizens. At just over three and a half minutes in, one of the officers radios for assistance and Luke shoots them both, exiting the car to fire further shots at point blank range. Theo follows him, asks him why he killed them, and Luke tells him to get back in the car, threatening him at gunpoint. When Theo gets back in the car, the camera remains at the scene, panning from the disappearing car to the dead police officers on the road, which is the final frame of the shot.

The second long take, which I have called the rescue, is 378, 379, or 390 seconds long, depending on which critic one reads. My opinion is that it is 378 seconds (6 minutes and 18 seconds) in length (83:52–90:10). It occurs at the end of the fourth act of the film, continues through the *all is lost moment* of monomythic storytelling, and concludes a third of the way through the final act. The single-shot sequence opens with a view of a tunnel as Theo, Kee, and Dylan, who are being escorted by Marichka (played by Oana Pellea) and Sirdjan (played by Faruk Pruti), attempt to escape from Bexhill. They hide in the tunnel as a gun battle between the Army and what appears to be a West African faction of refugees takes place in front of them. The West Africans flee, are pursued by the Army, and the five of them leave the tunnel and make for what seems to be an entrance to the sewers. Before Sirdjan gains access, Patric's voice is heard offscreen, telling them to put their hands up. He appears from the background with a handful of Fishes, punches Theo, and puts his pistol to his head. About a minute into the shot, Luke's voice is heard, also offscreen, telling Patric not to kill Theo. Luke appears with three more Fishes, rushes to Kee (who is in a wheelchair), and asks her how Dylan is. Luke and Theo have a brief conversation, then Luke wheels Kee and Dylan away and tells Patric to kill the other three as soon as he has left. The fourth act ends with Luke, Kee, and Dylan leaving, one and a half minutes into the shot. The final act begins with Patric shooting Sirdjan. He turns to shoot either Marichka or Theo next, but comes under fire. The Fishes advance to engage their assailants, who turn out to be the Army, and Theo and Marichka flee in the same direction as Luke, Kee, and Dylan. The Fishes retreat and Theo separates from Marichka. From this point onwards, the camera, whose focus has varied so far, remains close to Theo, either behind, alongside, or in front of him as he moves. Theo follows Patric into another street, but he reverses direction when the Army cut him off, sees Theo, and opens fire at him. Theo takes cover on an abandoned bus filled with other people seeking sanctuary. A few seconds after the third minute, two spots of blood appear on the camera's lens and remain there for just over a minute (suggesting that this is not an actual single-shot sequence either). The effect was accidental rather than planned, but is widely regarded as one of the highlights of the film's cinematography, encapsulating its aesthetic (Riesman 2022). Theo watches the Fishes take refuge in a big block of flats as armoured fighting vehicles approach and crosses the road, once again taking cover as the gunbattle intensifies, and the change of lighting shows several more spots of blood on the camera lens. At just over four minutes, Theo enters the flats, which are in ruins and filled with the dead and wounded. He takes the stairs to the first floor and finds Patric firing out the windows. Patric is shot, Theo continues his search for Kee, hears Dylan cry, and walks up to the second floor. Just under six minutes into the shot, he finds Kee and Dylan, cowering beneath a window ledge as Luke fires at the soldiers below. Theo reaches Kee, helps her up, and is about to leave the room when Luke sees him. The shot ends with a frame of Luke pointing his assault rifle at Theo.

Each of the long takes deserves a lengthy description because a more concise summary would communicate neither their technical expertise nor the impact of that technical expertise on both the narrative and the audience. Considering the ambush in isolation, the close, claustrophobic camerawork not only heightens the suspense and tension once the action begins, but also encourages the audience to sympathise, empathise, or identify with Theo. Thus far, he has been presented as a cold, cynical alcoholic who may only be helping the Fishes for financial reward. The initial conversation in the car presents a different side of him and makes his residual feelings for Julian clear. This is exacerbated when Julian is shot and Theo's extreme distress is evinced in immediate and visceral detail. Finally, the long take is one of only two times in the film that Theo uses violence, defending himself and Kee on both occasions and making his life meaningful by taking action rather than remaining passive. The ambush is also the first suggestion that *Children of Men* is not a typical action thriller with the early death of Julian, who was presented as Theo's romantic interest in the first two acts.

The rescue is both longer and slower than the ambush, with all of the action taking place on foot as Theo walks, runs, and hides. The close, shaky camerawork reproduces Theo's experience of the chaos in Bexhill for the audience, heightening the suspense and tension of the narrative while encouraging further sympathy, empathy, or identification with Theo. Here, his true heroism is presented as he literally runs into the middle of a battle to rescue Kee and Dylan while unarmed, undeterred by either Patric targeting him directly or the Army's targeting of all refugees, whether armed or unarmed. Although I have called the long take the rescue, it ends before Kee is rescued, with Luke holding Theo, Kee, and Dylan at gunpoint. When Luke takes fire from the Army, he returns it, Theo pushes Kee and Dylan out the room, and is shot by Luke as he leaves (though it looks like Luke has missed). Luke dies offscreen and the rescue succeeds when the soldiers allow Theo, Kee, and Dylan to walk through their ranks. Even if one knows little or nothing about cinematography, the two long takes are so intimate, intense, and impelling that one cannot help but recognise that there is something both different and significant about them. Their juxtaposition at the end of acts two and four invites comparison and demonstrates Theo's development, from a nostalgic cynic who loses everything that has meaning in his life (when Julian is killed and before he discovers Kee's pregnancy) to his commitment to a new cause (the future of humanity) for which he is prepared to (and does) sacrifice his own life. His death is a victory rather than defeat and he confronts it with calm and courage because he is convinced that the sacrifice is worthwhile and that it has restored meaning to his life. The juxtaposition of the two long takes with Theo's death is one of the ways in which his achievement is represented, revealed, and reiterated.

Notes

1 The blog, which has been archived and edited, remains available at the time of writing, at: <http://k-punk.abstractdynamics.org/>.

2 Fisher's (1998) article is worth reading for enthusiasts, but bears no relation to a conventional or critical criminological journal article in either content or form.

3 Further education, often abbreviated to FE (to distinguish it from HE, higher or tertiary education) refers to colleges (rather than secondary schools) that cater for 16- to 18-year-olds in the UK.

4 The actual origin of the phrase is H. Bruce Franklin (1979), in an essay on one of J.G. Ballard's apocalyptic narratives.

5 As is already obvious by this point, it is at times difficult to distinguish between the commentary and criticism of Fisher and Žižek. I do not think this is problematic, nor that one needs to identify who said what first – which would be difficult if not impossible given Fisher's initial use of a blog for his writing and the frequency with which Žižek sits for interviews – because each provides evidence for the other's position.

References

Arthur, C. (2012). The History of Smartphones: Timeline. 24 January. *The Guardian*. Available at: <www.theguardian.com/technology/2012/jan/24/smartphones-timeline>.

Bar-El, E. (2023). *How Slavoj Became Žižek: The Digital Making of a Public Intellectual*. Chicago: The University of Chicago Press.

Bielik, A. (2006). 'Children of Men': Invisible VFX for a Future in Decay. 27 December. *Animation World Network*. Available at: <www.awn.com/vfxworld/children-men-invisible-vfx-future-decay>.

Children of Men (2006). Directed by Alfonso Cuarón. US: Universal Pictures.

Children of Men – Comments by Slavoj Žižek (2007). In: *Children of Men – Special Edition*. Directed by Alfonso Cuarón. DVD, B000NIMZZM. Universal City, CA: Universal.

Fisher, M. (1998). White Magic. *The Red Feather Journal of Postmodern Criminology: An International Journal*, 6 (4). Available at: <https://critcrim.org/redfeather/journal-pomocrim/vol-6-virtual/whitemagic.htm>. Accessed 31.10.24.

Fisher, M. (2009). *Capitalist Realism: Is There No Alternative?* Alresford: Zero Books.

Fisher, M. (2014). *Ghosts of My Life: Writings on Depression, Hauntology and Lost Futures*. Alresford: Zero Books.

Fisher, M. (2016). *The Weird and the Eerie*. London: Repeater Books.

Fisher, M. (2018a). *Flatline Constructs: Gothic Materialism and Cybernetic Theory-Fiction*. New York: Exmilitary Press.

Fisher, M. (2018b). *k-punk: The Collected and Unpublished Writings of Mark Fisher (2004–2016)*. London: Repeater Books.

Fisher, M. (2022). *Capitalist Realism: Is There No Alternative?* 2nd ed. Alresford: Zero Books.

Franklin, H.B. (1979). What Are We to Make of J.G. Ballard's Apocalypse? In: Clareson, T.D. (ed.). *Voices for the Future: Essays on Major Science Fiction Writers: Volume Two*. Bowling Green, OH: Bowling Green University Popular Press, 82–105.

Fukuyama, F. (1992). *The End of History and the Last Man*. London: Penguin Books.

Goddard, T. (2021). Reacquiring Zero Books – a Statement. 27 November 2021. *Repeater Books*. Available at: <https://repeaterbooks.com/reacquiring-zero-books-a-statement/>. Accessed 31.10.24.

Hall, S. (2012). *Theorizing Crime and Deviance: A New Perspective*. London: Sage Publications.

Hall, S. & Winlow, S. (2015). *Revitalizing Criminological Theory: Towards a new Ultra-Realism*. Abingdon: Routledge.

Heidegger, M. (1927/1962). *Being and Time*. Trans. J. Macquarrie & E. Robinson. New York: Harper Collins Publishers.

IMDbPro (2025). United 93. *Box Office Mojo*. Available at: <www.boxofficemojo.com/releasegroup/gr2814464517/>.

Jacobson, G. (2020). Why Children of Men Haunts the Present Moment. 22 July. *The New Statesman*. Available at: <www.newstatesman.com/culture/film/2020/07/children-of-men-alfonso-cuaron-2006-apocalypse-coronavirus>.

James, P.D. (1992/2018). *The Children of Men*. London: Faber & Faber.

Jameson, F. (1981/2002). *The Political Unconscious: Narrative as a Socially Symbolic Act*. New York: Routledge.

Jameson, F. (1991). *Postmodernism, or, the Cultural Logic of Late Capitalism*. Durham, NC: Duke University Press.

Jameson, F. (2002). *A Singular Modernity: Essay on the Ontology of the Present*. London: Verso Books.

Jameson, F. (2005). *Archaeologies of the Future: The Desire Called Utopia and Other Science Fictions*. London: Verso Books.

Jameson, F. (2007). *The Modernist Papers*. London: Verso Books.

Jameson, F. (2013). *The Antimonies of Realism*. London: Verso Books.

Jameson, F. (2019). *Allegory and Ideology*. London: Verso Books.

Johnston, A. (2008). *Žižek's ontology: A Transcendental Materialist Theory of Subjectivity*. Evanston, IL: Northwestern University Press.

Martinez, R.R. (2019). First as Warning, Then as Threnody: On Cuarón's 'Children of Men'. 4 November. *PopMatters*. Available at: <www.popmatters.com/children-men-alfonso-cuaron-hegel>.

Ortiz-Ospina, E. (2019). *The Rise of Social Media: Our World in Data*. 18 September. Available at: <https://ourworldindata.org/rise-of-social-media>. Accessed 31.10.24.

The Possibility of Hope (2007). In: *Children of Men – Special Edition*. Directed by Alfonso Cuarón. DVD, B000NIMZZM. Universal City, CA: Universal.

Raymen, T. (2022). *The Enigma of Social Harm: The Problem of Liberalism*. Abingdon: Routledge.

Riesman, A. (2022). Why 'Children of Men' Remains Relevant with Each Passing Year. 18 November. *Special Broadcasting Service*. Available at: <www.sbs.com.au/whats-on/article/why-children-of-men-remains-relevant-with-each-passing-year/n3tnyumqr>.

Wolmart, G. (2017). On Anamorphic Adaptations and the Children of Men. *International Journal of Žižek Studies*, 11 (2), 1–21.

Žižek, S. (1989). *The Sublime Object of Ideology*. London: Verso Books.

Žižek, S. (1999a). *The Ticklish Subject: The Absent Centre of Political Ideology*. London: Verso Books.

Žižek, S. (1999b). Preface: Burning the Bridges. In: Wright, E. & Wright, E. (eds.). *The Žižek Reader*. Malden, MA: Blackwell Publishing, vii–x.

Žižek, S. (2006). *The Parallax View*. Cambridge, MA: MIT Press.

Žižek, S. (2008). *Violence: Six Sideways Reflections*. New York: Picador Books.

6 Symbolism

In my introduction to Chapter 4, I noted that Alfonso Cuarón's *Children of Men* (2006) seemed destined for obscurity by April 2007, following a disappointing performance at the Academy and BAFTA Awards and unremarkable DVD sales. I suggested reasons for its lack of commercial success in Chapter 5, summarised as conspicuously poor marketing by Universal Pictures and the residual racism and classism of Anglophone and other audiences. Given that the film had always been a critical success, it is perhaps unsurprising that it acquired the status of a 'cult classic' in the second decade of the 21st century, with its virtues revisited in *Sight & Sound*, *The Pop Culture Philosopher* blog, and *Rolling Stone*. There was a revival of public interest in 2015, with the European migrant crisis, which was sustained by Brexit (2016) in the United Kingdom, Donald Trump's presidency in the United States (2017–2021), and the global phenomenon often referred to as the rise of right-wing populism (Riesman 2022). This revival became exponential during the COVID-19 pandemic (2020–2022) and was briefly reinvigorated in the United Kingdom when plans to reopen Her Majesty's Prison Northeye, in Bexhill-on-Sea, as an immigration detention centre were announced in 2023. Gavin Jacobson (2020) is entirely accurate when he states: 'the film has since become the cultural exemplum of apocalypse, a singularly bleak imagining of our collective demise.'

6.1 Kee's Journey

In Chapter 2, I described Fredric Jameson's (1981, 1991, 2002, 2005, 2007, 2013, 2019) aesthetic education as composed of three interlocking concepts – capitalist realism, genuine allegory, and utopian desire. Genuine allegory, which I explained in Chapter 3, functions so as to reveal the structure of fourfold meaning within a text and in so doing creates meaning in excess of the representational capacity of the text. Genuine allegories are fourfold and thick and stage the complexity of social reality by holding up both a mirror and a microscope to everyday life (Jameson 2019). The literal level of meaning of an allegory is the meaning of the represented sequence of events.

DOI: 10.4324/9781003565826-8

I applied Joseph Campbell's (2008) summary of the monomyth as a summary of *Children of Men* in Chapter 4:

> Theo ventures forth from the world of physical and spiritual infertility into a region of revolutionary hope and commitment: Luke and the Fishes are there encountered and a decisive victory is won when Theo helps Kee escape their clutches: Theo comes back to humanity with the power to save Kee and her newborn child.

The literal level is where the two superstructures, monomyth and allegory, intersect. My preference is to understand the plot (and, indeed, narrative structure) as an essentially ethical concept, in which case an alternative but completely compatible summary of *Children of Men* is (as also described in Chapter 4):

> the movement from an inaugural condition in which a cynical, dispassionate Theo leads a meaningless life (*is-but-ought-not-to-be*) to his decision to dedicate himself to Kee's protection when he discovers her pregnancy to a retrospectively inevitable condition in which Theo makes his life meaningful by sacrificing it for the future of humanity (*is-and-ought-to-be*).

These descriptions are not erroneous, but they do miss something significant about the film. In the previous two chapters, I have repeatedly referred to the subtlety of the narrative, a subtlety only revealed in full on repeat viewings when one realises how much takes place offscreen, in the background, or hidden in plain sight. It is not simply a matter of quantity but quality too. In my summary of *Children of Men* in terms of John Yorke's (2013) template in Chapter 4, for example, I described the call to action at the end of the first act. Theo initially fails to answer this call by turning Julian and the Fishes down. Somewhere – or, more accurately, *somewhen* – in between the end of the first act and beginning of the second act, Theo changes his mind (or perhaps does not change his mind if he was lying to Julian). The acceptance of the call to action is an unusually significant (and dramatic) moment to leave to the imagination of the audience. Consider next the turning point. The 'turn' is the change in Theo's motivation from what he wants (an easy, idle life, the best part of which is over) to what he needs (a new cause to which he can commit and which will return meaning to his life). But Theo does not *do* anything in the sequence in which this turn occurs: Kee initiates the conversation while he is watching Miriam practise some kind of Tai Chi, and all he does is respond. It is so subtle that it took me many viewings to understand that Cuarón does actually employ monomythic storytelling.

Similarly, Theo dies *before* the Human Project boat arrives, leaving a remarkably calm and courageous Kee with her newborn baby, bobbing on a none-too-seaworthy rowing boat, in a thick mist. She reaches for Theo when she sees the boat, but he has been dead for at least half a minute of screen time, perhaps longer. Yes, Kee and Dylan are only there because they were first rescued from the Fishes (in the second long take I discussed in Chapter 5) and then rowed by Theo, but he does not live long enough to know whether the mother and child will make the rendezvous or suffer a fate even worse than that planned by the Fishes (either drowning or being enslaved and exploited by someone else). Theo's transformation from nostalgic cynic to self-sacrificing protector is certainly a personal victory, but it is highly unconventional to deny him at least a glimpse of the fruits of that victory before he dies. Bearing in mind that he has always been sceptical about the existence of the Human Project, he dies with nothing but hope: hope that the Human Project exists, hope that the information about the rendezvous is accurate, and hope that nothing has happened to prevent the Human Project from reaching the rendezvous. As such, from Theo's point of view, his success is limited, saving Kee and Dylan from one harm while placing them in the way of others.

My suggestion is that while Cuarón does indeed employ monomythic storytelling, Kee is the protagonist and *Children of Men* her story rather than Theo's. In other words, the hero's journey is:

> Kee ventures forth from the world of physical and spiritual infertility into a region of propagation and hope: Luke and the Fishes are there encountered and a decisive victory is won when she gives birth to Dylan and escapes their clutches: Kee comes back to the world with a newborn child and the power to save humanity from extinction.

As with the superficial version in which Theo is the protagonist, several key elements of this journey take place offscreen (in Kee's case, prior to the start of the narrative). Her call to action is the discovery that she is pregnant, her response is her decision to keep the baby, and her initial objective is achieved when she places herself under Julian's protection. The turning point remains the same (but is a product of Kee's agency as she initiates friendship with Theo), all is lost when she is captured by Luke, and there is resolution in victory when she sees the Human Project boat arrive. This might seem a counter-intuitive interpretation of *Children of Men* as an instantiation of the monomyth, but it has much greater explanatory power than the more obvious analysis. Theo is, after all, an everyman, lacking in any remarkable qualities and entirely insignificant to the overarching story of human extinction – with the exception of the role he plays in rescuing Kee and Dylan. Recall Slavoj Žižek's (cited in *Children of Men* 2007, 00:07–00:39, emphasis in original) commentary on the film from Chapter 5:

> I would say that the true focus of the film is there in the background and it's crucial to leave it *as* a background. Here comes his true art – Cuarón's – it's

> the paradox of what I would call an anamorphosis. If you look at the thing too directly . . . you don't see it.

Kee is the background and Cuarón brings her journey into a sharper focus than might otherwise be achieved by using Theo and his journey as a prism.

The symbolic level is the hidden or secret meaning of the represented sequence of events. One way to think about the symbolic level is that in spite of being hidden or secret, it establishes a direct and robust relationship between the work (characters, settings, and action) and the world in which that work is produced, consumed, and experienced (people, places, and events). I have previously argued, for example, for the following set of relations between representation and reality in popular cultural allegories: sex is gender (*Mad Max: Fury Road*), speciesism is racism (*Carnival Row*), and the Army of the Dead is ecocide (*Game of Thrones*) (McGregor 2021, 2023a). In answer to a question about *Children of Men's* relevance to current society, Francis Fukuyama (2016: 3:43–4:48) makes the following claim:

> This is in a way the scenario that is now being played out in Japan and South Korea and Taiwan and a number of really low fertility countries where the population is basically shrinking and aging and actually doing that very rapidly – not, obviously, as it happens in *Children of Men*, but, you know, the idea that we are constantly going to grow, you know, that next year's going to be bigger, there'll be more buildings and more, you know, stuff to do, and more technologies is actually, you know, in . . . in certain parts of the world, all of that's going into reverse and so there aren't, you know, new schools opening and . . . and . . . and new stuff happening and in fact people are consolidating, they're leaving buildings that they've lived in for generations and . . . and . . . and, you know, and so that . . . that the movie really does speak to, you know, a real social situation I think that many countries are probably going to face in the future.

For Fukuyama, the fertility crisis in *Children of Men* symbolises the aging populations of not only the countries he mentions, but the global population.[1]

The symbolic meaning of an allegory is always more subtle and complex than the literal meaning and given the subtlety and complexity of the literal meaning of *Children of Men*, Fukuyama's interpretation is an oversimplification. While I agree with him about the primary locus (or focus) of the symbolism, in or on the fertility crisis, the crisis in the film is (like the Army of the Dead in *Game of Thrones*) the climate crisis in reality, the first of the three aporias of capitalist realism described by Mark Fisher (2022).[2] I shall present evidence throughout this chapter, but my starting point is that the fertility crisis in *Children of Men* is the triumph of nature over culture, the planet fighting back to save itself from the destruction wreaked by humanity. Like the real climate crisis, the fictional fertility crisis is anthropocidal, that is threatens the existence of the entire human species. Like the climate crisis, the fertility crisis is also a slow apocalypse, whose impact increases gradually and steadily rather than dramatically (although the fertility crisis in the film will destroy the species far quicker than the climate crisis in reality). Finally, the fertility crisis is also, like the climate crisis, a certainty: if nothing changes (the discovery of a cure in the film), then anthropocide will occur. If this interpretation seems to stretch credibility, consider once again Žižek's anamorphic reading of the film in which the background is more important than the foreground, whose primary function is to hold a microscope to the background. The fertility crisis is a crisis of too few people, whereas the climate crisis is a crisis of too many people (actually, of too much consumption, but consumption is supervenient on population). The falling fictional population is the foreground and the rising real population the background.

There is another reason why Cuarón may have opted to place the climate crisis in the background, which Fisher (2022: 18) suggests when he explains environmental catastrophe in terms of Jacques Lacan's (1973, 1986, 2005) register of the Real, which refers to the ineffable world, only detectable by human beings indirectly through the unconscious:

> For Lacan, the Real is what any 'reality' must suppress; indeed, reality constitutes itself through just this repression. The Real is an unrepresentable X, a traumatic void that can only be glimpsed in the fractures and inconsistencies in the field of apparent reality. So one strategy against capitalist realism could involve invoking the Real(s) underlying the reality that capitalism presents to us. Environmental catastrophe is one such Real. At one level, to be sure, it might look as if Green issues are very far from being 'unrepresentable voids' for capitalist culture. Climate change and the threat of resource-depletion are not being repressed so much as incorporated into advertising and marketing. What this treatment of environmental catastrophe illustrates is the fantasy structure on which capitalist realism depends: a presupposition that resources are infinite, that the earth itself is merely a husk which capital can at a certain point slough off like a used skin, and that any problem can be solved by the market.

The climate crisis is the Real that capitalist civilisation has suppressed and that has, according to Jason Moore (2015), shifted capitalist world-ecology to negative value and will, in the not-too-distant future, cause the collapse of global capitalism. Fisher also alludes to the fact that climate change is extremely difficult to represent in any meaningful way, which is a consequence of its scale and complexity, among other features. If he is right – and it seems clear to me that he is – then it is precisely the type of phenomenon that one might choose to represent symbolically, to leave as a background that is accessed by and through the foreground in the same way that the Real is glimpsed through the fractures and inconsistencies of the unconscious. The fertility crisis is the microscope through which Cuarón, paradoxically and perhaps even wryly, examines the climate crisis in *Children of Men*.

6.2 Late Capitalist Logic

The existential level is the meaning of individual desire and the construction of subjectivity in the represented sequence of events. In the previous section, I claimed that the fertility crisis in *Children of Men* symbolises the climate crisis in reality and made several comparisons between the two. One of the contrasts is that there are no fertility crisis deniers. The fertility crisis is much easier to understand than the climate crisis and is happening much more quickly. One cannot in good faith deny infertility because a quick look around reveals the absence of children and it is a simple extrapolation from the present absence to future anthropocide: of those already born, the last to die – in 30, 50, or 80 years' time – will be the last human being ever.[3] The existential meaning of *Children of Men* is thus focused on the construction of subjectivity in the face of anthropocide. For Martin Heidegger (1927), all human lives are shaped by and lived in response to our awareness of our mortality, but in the film that awareness is of both our own mortality and the knowledge that each human death brings the species one step closer to extinction. In Chapter 1, I stated that ecocide is not genuinely anthropocidal because human history suggests that we will kill one another long before we destroy the Earth's capacity to support human life (and such decivilisation might well prevent the complete destruction of the planet's capacity to support human life). While anthropocide is happening much more quickly in *Children of Men* than in reality, the characters are nonetheless faced with decivilisation and technological decline (there are, for example, no mobile phones in 2027 United Kingdom even though they were ubiquitous in the United Kingdom during the making of the film). Living one's life and constructing one's subjectivity in such circumstances is one of the two features of the film that attracted Fukuyama (2016) to it. He regards *Children of Men* as posing the question of whether and how knowledge of a distant but certain anthropocide would affect individuals' decisions about their behaviour and future plans: for example, would they live lives similar to those we live now or would social order simply break down?

As discussed in Chapter 4, Žižek (*Children of Men* 2007; *The Possibility of Hope* 2007) asks a similar question in terms of Cuarón's thematic exploration of the conflict between hope and despair. What is perhaps most interesting about this theme is that Cuarón clearly means it to apply to the world in which we watch *Children of Men* as well as the world of the film because the documentary *The Possibility of Hope* is about whether hope is a plausible response to the reality not the fiction. The fiction is, in consequence, a comment on the reality and a study in whether and how hope can give meaning to one's life in even very extreme circumstances. Theo's psychological journey, from cynicism and despair to hope and commitment, stages this conflict and provides the audience with Cuarón's answer to his own question. In spite of his premature death and in spite of that death occurring before he can be certain of success, Theo's life has undoubtedly been enriched by his dedication to Kee and Dylan and the renewal of human relationships that dedication involved. His life is not only enriched and meaningful, but he actually dies happy, his (psychological) well-being peaking at his (physical) death. Žižek (*Children of Men* 2007: 05:10–05:26, emphasis in original) makes much of this scene in the film:

> What I like is that the solution is the boat. It doesn't have roots. It's rootless. It floats around. This is for me the meaning of this wonderful metaphor: *boat*. The condition of the renewal means you cut your roots. That's the solution.

This may seem to contradict what Žižek says about *true infertility as the lack of meaningful historical experience*, but renewal does not require a rejection of history, only a commitment to a new form of life or structure of feeling. If the boat is the collective ideology within which human subjectivity is constructed, then that construction is essentially rather than accidentally hopeful: in the boat, one is both rootless and at the mercy of the elements. In the boat, one might die (like Theo) or be rescued (like Kee and Dylan) and one just has to hope for the best rather than cling on to the old and outdated.

Žižek (2008) subsequently discusses the meaning of *Children of Men* in relation to Friedrich Nietzsche's (1883) last man, a subject picked up by Fukuyama (1992) in *The End of History and the Last Man.* As mentioned in Chapter 5, Žižek (2008: 29) describes the last man as 'an apathetic creature with no great passion or commitment. Unable to dream, tired of life, he takes no risks, seeking only comfort and security, an expression of tolerance with one another.' The last man is the product of the end of history, by which Fukuyama means not a termination of temporality (which would be inconceivable), but the end of a human history driven by conflict between what G.W.F. Hegel (1807) referred to as master and slave. The end of history is in fact a great human achievement because it entails the end of war and other mass harms. The last man is, however, in Nietzsche's controversial conception, an inferior being – precisely because his (or *her* – the last man is really the last *person*) life is too comfortable and because he has nothing left to fight for and no risk of death to make his life meaningful. Fukuyama argued that the end of history had been reached with the triumph of liberal democracy and the market economy over totalitarian communism and the planned economy at the end of the Cold War. His (Fukuyama 1992: 312) idea was that it was only a matter of time before liberalism and capitalism achieved global acceptance and turned the world's population into the last men:

> The life of the last man is one of physical security and material plenty, precisely what Western politicians are fond of promising their electorates.

> Is this really what human history has been 'all about' these past few millennia? Should we fear that we will be both happy and satisfied with our situation, no longer human beings but animals of the genus *homo sapiens*?

Both Nietzsche's and Fukuyama's conceptions are problematic. To begin with, they make an unsubstantiated link between conflict and civilisation and ignore the difficulty and challenge of sustaining a democracy in which there is genuine social justice. Fukuyama is also Eurocentric, either ignoring China or assuming it will follow Western suit, which seems highly unlikely three decades later. Fukuyama furthermore failed to anticipate how much inequality and polarisation would occur in the new, neoliberal world order, particularly after the Great Recession (2007–2009).

Žižek interprets *Children of Men* as representing the populations of democracies in the Global North as Nietzsche's last men. There is of course a sense in which this is accurate. To begin with, the characters in the film are the literal rather than metaphorical last people. Theo, for one, exhibits some of the vices of the last man before his psychological transformation. He complains to Jasper about being numb, but then drinks alcohol and smokes cannabis to maintain that numbness. One might even say that he is *an apathetic creature with no great passion or commitment who is unable to dream, is tired of life, takes no risks, and seeks only comfort and security*. One might then say that such a life is a life lived in despair rather than hope and that, as mentioned before, Theo's psychological transformation is a movement from despair to hope. The problem with Žižek's interpretation of Theo as the last man is that it is inconsistent with the setting of *Children of Men*: his world is not a civilisation at the end of history, with abundance, prosperity, and peace, but a world in which civilisations have collapsed and are continuing to collapse, characterised by conflict and chaos and living in the face of anthropocide. Even in the relative safety of the United Kingdom, the population are not last men, but people who are living in and through an apocalypse. Similarly, extrapolating from the fiction to reality, the circumstances of the last man in the 21st century are very different from that predicted by either Nietzsche or Fukuyama and the concept is perhaps most useful for the extent to which it is inaccurate and inapplicable.

The anthropic level is the meaning of human achievement and collective ideology in the represented sequence of events. Existential meaning is embedded in anthropic meaning. I described the existential meaning of *Children of Men* as concerned with the construction of subjectivity in the face of – *towards* – anthropocide and the question of whether the individual should live towards anthropocide with either hope or despair. This exploration of the conflict between hope and despair is situated within a narrative of collective ideology and the ideology dramatised by *Children of Men* is most accurately described by Fisher (2022: 2) in a passage I cited in Chapter 5, but is worth repeating here: 'ultra-authoritarianism and Capital are by no means incompatible: internment camps and franchise coffee bars co-exist.' The compatibility of capitalist world-ecology with mass harms is evident from the very beginning of capitalist civilisation, the second half of the 15th century according to Moore (2015), and has included genocide, slavery, colonialism, and ecocide as well as classism, racism, and sexism (Robinson 1983; Salleh 2017; Andrews 2021). More recently, as discussed in Chapter 3, neoliberal capitalism has been compatible with a change of criminal justice culture from an ideal of social progress to an ideal of zero tolerance, which has been used to justify a security state based on the twin pillars of market discipline and moral discipline. In the 21st century, the security state has been expanded and emboldened by two decades of the Global War on Terrorism (Garland 2001; Wacquant 2004; Žižek 2008).

Pierre Dardot and Christian Laval (2016: xx) argue that both market and moral discipline were aggravated rather than alleviated by the Great Recession and refer to post-crash neoliberalism as the *new neoliberalism*, which they link to right-wing populism and summarise as a 'war against the population'. New neoliberalism is identical with or very close to the setting of *Children of Men*, which Žižek (2008: 28) describes as follows:

> The U.K. lives in a permanent state of emergency: antiterrorist squads chase illegal immigrants, the state power administering a dwindling population which vegetates in sterile hedonism. Hedonist permissiveness plus new forms of social apartheid and control based on fear – are these not what our societies are now about? But here is Cuarón's stroke of genius: 'Many of the stories of the future involve something like "Big Brother," but I think that's a twentieth-century view of tyranny. The tyranny happening now is taking new disguises – the tyranny of the twenty-first century is called "democracy."'

Žižek's description echoes Fukuyama's (2016) first reason for selecting the film as one of his favourites, which is that it depicts a potential future for the United Kingdom following Brexit and the rise of the right. In citing Cuarón, Žižek reveals another flaw in Fukuyama's end of history: the coincidence of the joint triumph of liberal democracy and the market economy at the end of the 20th century was precisely that, a coincidence, because capitalism is also compatible with totalitarianism. This should have been apparent to Fukuyama, even at his time of writing, because the People's Republic of China was one of the drivers of the neoliberal revolution, along with the United States.

Fukuyama (2018) has admitted that this joint victory was not as benign as he predicted, blaming the steep increase in economic inequality on factors he failed to identify. His report of a revolution in 1992 was accurate, but he misidentified both its timing and nature. The revolution was economic not political, the end of politics rather than the end of history, and had begun at least a decade earlier. Stuart Hall (2011: 705) describes the 'long march of the

Neo-Liberal Revolution' in the United Kingdom as having begun prior to the globalisation of 1978 to 1980 and continuing in various guises – Thatcherism, Blairism, and the many Conservative and coalition governments from 2010 to 2024 – for five decades. His definition is similar to David Harvey's (2005) in describing the leading idea of neoliberalism as being that the state must never interfere with the individual's right to make profit and amass wealth. This reduced 'function of the liberal state should be limited to safeguarding the conditions in which profitable competition can be pursued without engendering Hobbes' "war of all against all"' (Hall 2011: 707). The real revolution was thus not Fukuyama's liberalism, but Hall's neoliberalism and if the former can be characterised by its emphasis on the freedom of the individual, the latter can be characterised by its emphasis on competition as the primary motivation of the free individual. Fukuyama's victory of liberal democracy and the market economy and Hall's neoliberal revolution are identical with Fisher's capitalist realism.

Fisher (2022: 8) describes the essence of capitalist realism as follows: 'The 80s were the period when capitalist realism was fought for and established, when Margaret Thatcher's doctrine that "there is no alternative" – as succinct a slogan of capitalist realism as you could hope for – became a brutally self-fulfilling prophecy.' This is why I described the neoliberal revolution as the end of politics rather than the end of history. The neoliberal revolution was the widespread acceptance that there is no (longer any) alternative to the market economy, for which the end of the Cold War provided apparently irrefutable evidence when interpreted as the global victory of liberal democracy over totalitarian communism. Fisher acknowledges his debt to both Fukuyama and Jameson while contrasting the former's approval of capitalist realism with the latter's disapproval of what he refers to as the *cultural logic of late capitalism*. Like Jameson, Fisher (2022: 7) argues that it is not only politics that has been reduced to economics but culture as well, which has 'become dominated by pastiche and revivalism', as predicted by Jameson (1991). Capitalist realism or 'Really Existing Capitalism' means both political and cultural sterility and it is precisely such a society that is represented in *Children of Men* (Fisher 2022: 45). This is the most useful way to conceive of Nietzsche's last man in contemporary terms, as *homo economicus* locked in the political and cultural sterility of capitalist realism (Ailon 2020). As noted above, the sterility in *Children of Men* is both metaphorical, reproducing the reality of capitalist realism onscreen, and literal, in the infertility pandemic that underpins the plot. Fisher (2018: 54) makes an important point about the experience of capitalist realism that dovetails neatly with his initial definition and offers further evidence for my claim that the last man is *homo economicus*:

> Capitalist realism can be described as the belief that there is no alternative to capitalism. However, it is more usually manifest not in grand claims

> about political economy, but in more banal behaviours and expectations, such as our weary acceptance that pay and conditions will stagnate or deteriorate.

Capitalist realism is the collective ideology within which the existential exploration of the conflict between hope and despair is situated. Capitalist realism, as represented in the fiction and as a reflection of the reality in which we watch it is, has succeeded to the extent that it is 'coming close to achieving the ultimate goal of ideology – invisibility' (Fisher 2018: 462).

6.3 Being-Towards-Ecocide

In order to articulate *Children of Men's* unique contribution to green cultural criminology, one must move beyond its representational capacity, approaching the film as an event rather than a text (or, more accurately, *in addition to* a text). The narrative event is an allegorical staging of the complexity of social reality that discloses the relationships among the harms of ecocide, despair, and capitalist realism by the combination of character, setting, and action. Approaching *Children of Men* as an event determines and realises that event as a replication of *being-towards-ecocide*, that is lived experience of the climate crisis (Burns 2023). The event is extra-representational because the allegory does not only represent the reality of living in and through the climate crisis, but replicates the experience of the audience, who are living in and through the climate crisis as they experience the representation of that crisis onscreen.[4] The staging of being-towards-ecocide is a consequence of the thickness of genuine allegories, which determines and realises unpredictable, unexpected, and unusual causes and effects of the complexity of social relations. The idea that there is something different, perhaps even unique, about the way in which the climate crisis is presented in *Children of Men* is originally from Fisher (2022: 2–3) and worth quoting in full:

> The catastrophe in *Children of Men* is neither waiting down the road, nor has it already happened. Rather, it is being lived through. There is no punctual moment of disaster; the world doesn't end with a bang, it winks out, unravels, gradually falls apart. What caused the catastrophe to occur, who knows; its cause lies long in the past, so absolutely detached from the present as to seem like the caprice of a malign being: a negative miracle, a malediction which no penitence can ameliorate. Such a blight can only be eased by an intervention that can no more be anticipated than was the onset of the curse in the first place. Action is pointless; only senseless hope makes sense. Superstition and religion, the first resorts of the helpless, proliferate. But what of the catastrophe itself? It is evident that the theme of sterility must be read metaphorically, as the displacement of another

> kind of anxiety. I want to argue this anxiety cries out to be read in cultural terms, and the question the film poses is: how long can a culture persist without the new? What happens if the young are no longer capable of producing surprises? *Children of Men* connects with the suspicion that the end has already come, the thought that it could well be the case that the future harbors only reiteration and re-permutation.

Three key consequences follow from Fisher's core phrase *being lived through*, all of which are concerned with the temporality of the apocalypse. First, there is no *punctual moment*, no clear beginning, which makes it difficult to grasp. Second, the absence of a starting point means that the causes of the apocalypse are *absolutely detached* from everyday life in the present, which obfuscates meaningful mitigatory action. Finally, there is *the suspicion that the end has already come*, that it is too late to take action or that one is already living in a post-apocalyptic era. All three of these features of the fictional fertility crisis are features of the climate crisis and are in fact more relevant to the latter than the former given that it is happening more slowly and has more complex causes. Rudy Ralph Martinez (2019) concurs, describing *Children of Men* as a representation of 'living through the apocalypse' and Gavin Jacobson (2020) brings this concept back to Fisher's capitalist realism:

> But more terrifying, perhaps, is how, even as the world's end approaches, we are still forced to endure the banalities of daily life. In other films about the apocalypse, such as *Mad Max*, global ruin invites a kind of dark liberation, when survivors embrace sex, drugs, racing, fighting and looting. But in *Children of Men*, Britain limps on and its citizens limp on in it. Humanity has reached its tragic denouement, but no one can escape the ordinary unhappiness of capitalist realism: endless forms, long commutes, idle queues, crowded streets, and trying to convince your boss to let you work from home.

Is this replication of the climate crisis unique? Perhaps not, but it is certainly very rare. In a short essay on climate change cinema, I analysed four exemplary films representative of the genre (McGregor 2023b): Kevin Reynolds' *Waterworld*, Roland Emmerich's *The Day After Tomorrow*, George Miller's *Mad Max: Fury Road*, and Adam McKay's *Don't Look Up*. My initial categorisation was whether the apocalypse is part of the story's setting, exploring themes of adaptation and recovery, or a diegetic event, exploring themes of anticipation and mitigation. I then subdivided the first category into whether the apocalypse is explicitly represented as being anthropogenic (*Mad Max*) or not (*Waterworld*). I subdivided the impending apocalypse category into the speed with which it occurs and whether there is time to take preventative action (*Don't Look Up*) or not (*The Day After Tomorrow*). The vast majority of feature films can be characterised by at least one of these four subcategories, but none apply to *Children of Men*: neither the cause nor the date of the fertility crisis is revealed. All the audience is told is that the last woman to give birth did so in 2009 and Miriam's recounting of her experience as a midwife suggests that the crisis had already been in progress for some time. There are only two other works that represent living through the apocalypse with similar success, Octavia Butler's (1993, 1998) Earthseed series, consisting of the novels *Parable of the Sower* and *Parable of the Talents*, and HBO's *Game of Thrones* (2011–2019) television series. Notwithstanding, *Children of Men* may well be unique as a feature film, achieving in 100 minutes what the series do in approximately 27 hours of reading and 70 hours of watching respectively.

One might accept what I have said about replicating the apocalypse in progress, but then ask *so what?* Why or how is replicating everyday experience of the climate crisis valuable? In introducing the extra-representational meaning of *Children of Men*, I used the phrase 'being-towards-ecocide.' The term is, as suggested before, derived from Heidegger's (1927) analytic of being-here (often left untranslated as *Dasein*) in *Being and Time*. It is, however, from Reece Burns' (2023) sophisticated exploration of the value of Heidegger's hermeneutic phenomenology for green criminology, in which he evaluates the analytic of being-here for the 21st century, that is for human agency in the face of (or living towards) the certainty of its own death *and* the likelihood of climate-change-induced societal collapse. For Heidegger, *being-towards-death* is not simply the agent's awareness of their mortality, but a particular comportment in which the agent neither avoids nor obsesses over her mortality, accepting, in Burns' (2023: 1143, emphasis in original) terms that she is '*being-ever-at-the-point-of-death.*' Burns argues that being-towards-death is an insufficient characterisation of human agency in the 21st century in consequence of the causal relation between unreflective, unselfconscious everyday life on the one hand and anthropocidal ecocide on the other. Reducing or preventing ecocide will require a transformation of familiar forms of life or structures of feeling, that is *being-towards-ecocide*, which Burns (2023: 46) defines as: 'an explicit way of existing that requires a re-attunement of the self and the complete overhaul of the individual human experience with respect to the annihilation of our species and the earth from which we ultimately derive.'

Heidegger maintains that being-towards-death is the only authentic mode of being, but is typically avoided by human agency in consequence of its uncanny, distressing, and dreadful character. In a similar manner and for almost exactly the same reasons, human agency typically avoids being-towards-ecocide by existing in an everyday mode focused on the present and short term. Being-towards-ecocide is precisely what is replicated by *Children of Men* in the unique (or at least extremely rare) way in which it establishes and sustains a relation between the lived experience of the represented apocalypse onscreen and the lived experience of the actual apocalypse in reality. Burns (2023: 1148) expands on being-towards-ecocide with the concepts of 'being-oneself-in-the-face-of-ecocide' and 'reckoning-with-ecocide,' and although this expansion is beyond the scope of my inquiry in this essay, it demonstrates the usefulness of and necessity for a reconception of forms of life or structures of feeling in relation to ecocide. With being-towards-ecocide explained, one can see how *Children of Men* explores the phenomenology of the apocalypse in terms of the relations among ecocide, despair, and capitalist realism, holding a mirror and a microscope to the climate crisis as a cultural crisis. To return to Jameson's aesthetic education, *Children of Men* provides new ways of thinking about the world in the combination of its form and style, reconceptualising climate catastrophe as a crisis that is already well underway rather than on the horizon.

Notes

1 The United Nations (2025) predicts that by 2050 there will be almost the same number of people aged 65 and over as aged 11 and under and that the former category will be twice the number of children aged 4 and under. While Japan does indeed have the world's oldest population (with over 28 per cent aged 65 and over), the problem is more prevalent in Europe than Asia: Italy, Finland, Portugal, Greece, Germany, Bulgaria, Croatia, France, Latvia, and Serbia all have populations older than South Korea and Taiwan, with over a fifth of their respective populations aged 65 and over (Population Reference Bureau 2020).
2 See also Martinez (2019).
3 One might almost say the same for the climate crisis. For a discussion of the visible evidence and the speed with which the climate is changing, see Wallace-Wells (2019).
4 For a discussion of the difference between representation on the one hand and reproduction, replication, and representing on the other, see McGregor (2026).

References

Ailon, G. (2020). The Phenomenology of *Homo Economicus*. *Sociological Theory*, 38 (1), 36–50.

Andrews, K. (2021). *The New Age of Empire: How Racism and Colonialism Still Rule the World*. London: Allen Lane.

Burns, R. (2023). From Meaning to Ecocide: The Value of Phenomenology for Green Criminology. *Critical Criminology: An International Journal*, 31 (4), 1137–1154.

Butler, O.E. (1993/2019). *Parable of the Sower*. New York: Grand Central Publishing.

Butler, O.E. (1998/2019). *Parable of the Talents*. London: Headline Publishing Group.

Campbell, J. (2008). *The Hero with a Thousand Faces*. 3rd ed. Novato, CA: New World Library.

Children of Men (2006). Directed by Alfonso Cuarón. US: Universal Pictures.

Children of Men – Comments by Slavoj Žižek (2007). In: *Children of Men – Special Edition*. Directed by Alfonso Cuarón. DVD, B000NIMZZM. Universal City, CA: Universal.

Dardot, P. & Laval, C. (2016/2019). *Never-Ending Nightmare: The Neoliberal Assault on Democracy*. Trans. G. Elliott. London: Verso Books.

Fisher, M. (2018). *k-punk: The Collected and Unpublished Writings of Mark Fisher (2004–2016)*. London: Repeater Books.

Fisher, M. (2022). *Capitalist Realism: Is There No Alternative?* 2nd ed. Alresford: Zero Books.

Fukuyama, F. (1992). *The End of History and the Last Man*. London: Penguin Books.

Fukuyama, F. (2016). My Favorite Movie with Francis Fukuyama: *Children of Men. Future Tense*. Washington, DC: New America. Available at: <www.youtube.com/watch?v=03SrMeLmOUc>. Accessed 31.01.24.

Fukuyama, F. (2018). *Identity: Contemporary Identity Politics and the Struggle for Recognition*. London: Profile Books.

Game of Thrones (2011–2019). Originally released 17 April. US: HBO.

Garland, D. (2001). *The Culture of Control: Crime and Social Order in Contemporary Society*. Oxford: Oxford University Press.

Hall, S. (2011). The Neo-Liberal Revolution. *Cultural Studies*, 25 (6), 705–728.

Harvey, D. (2005). *A Brief History of Neoliberalism*. Oxford: Oxford University Press.

Hegel, G.W.F. (1807/1976). *Phenomenology of Spirit*. Trans. A.V. Miller. New York: Oxford University Press.

Heidegger, M. (1927/1962). *Being and Time*. Trans. J. Macquarrie & E. Robinson. New York: Harper & Row.

Jacobson, G. (2020). Why Children of Men haunts the present moment. 22 July. *The New Statesman*. Available at: <www.newstatesman.com/culture/film/2020/07/children-of-men-alfonso-cuaron-2006-apocalypse-coronavirus>.

Jameson, F. (1981/2002). *The Political Unconscious: Narrative as a Socially Symbolic Act*. New York: Routledge.

Jameson, F. (1991). *Postmodernism, or, the Cultural Logic of Late Capitalism*. Durham, NC: Duke University Press.

Jameson, F. (2002). *A Singular Modernity: Essay on the Ontology of the Present*. London: Verso Books.

Jameson, F. (2005). *Archaeologies of the Future: The Desire Called Utopia and Other Science Fictions*. London: Verso Books.

Jameson, F. (2007). *The Modernist Papers*. London: Verso Books.

Jameson, F. (2013). *The Antimonies of Realism*. London: Verso Books.

Jameson, F. (2019). *Allegory and Ideology*. London: Verso Books.

Lacan, J. (1973/1998). *The Seminar of Jacques Lacan Book XI: The Four Fundamental Concepts of Psychoanalysis*. Trans. A Sheridan. New York: W.W. Norton & Company.

Lacan, J. (1986/1992). *The Seminar of Jacques Lacan Book VII: The Ethics of Psychoanalysis 1959–1960*. Trans. D. Porter. New York: W.W. Norton & Company.

Lacan, J. (2005/2016). *The Sinthome: The Seminar of Jacques Lacan Book XXIII*. Trans. A.R. Price. Cambridge: Polity Press.

Martinez, R.R. (2019). First as Warning, Then as Threnody: On Cuarón's 'Children of Men'. 4 November. *PopMatters*. Available at: <www.popmatters.com/children-men-alfonso-cuaron-hegel>.

McGregor, R. (2021). *Critical Criminology and Literary Criticism*. Bristol: Bristol University Press.

McGregor, R. (2023a). *Literary Theory and Criminology*. Abingdon: Routledge.

McGregor, R. (2023b). The World-Ecology of Climate Change Cinema. *Theaker's Quarterly Fiction*, 75, 63–71.

McGregor, R. (2026). *Reducing Political Violence: Narrative, Critique, and Criminology*. Bristol: Bristol University Press.

Moore, J. (2015). *Capitalism in the Web of Life: Ecology and the Accumulation of Capital*. New York: Verso Books.

Nietzsche, F. (1883/2006). *Thus Spoke Zarathustra: A Book for All and None*. Trans. A. del Caro. Cambridge: Cambridge University Press.

Population Reference Bureau (2020). *Countries With the Oldest Populations in the World*. 23 March. Available at: <www.prb.org/resources/countries-with-the-oldest-populations-in-the-world/>.

The Possibility of Hope (2007). In: *Children of Men – Special Edition*. Directed by Alfonso Cuarón. DVD, B000NIMZZM. Universal City, CA: Universal.

Riesman, A. (2022). Why 'Children of Men' Remains Relevant with Each Passing Year. 18 November. *Special Broadcasting Service*. Available at: <www.sbs.com.au/whats-on/article/why-children-of-men-remains-relevant-with-each-passing-year/n3tnyumqr>.

Robinson, C. (1983/2000). *Black Marxism: The Making of the Black Radical Tradition*. Chapel Hill, NC: The University of North Carolina Press.

Salleh, A. (2017). *Ecofeminism as Politics: Nature, Marx and the Postmodern*. 2nd ed. London: Zed Books.

United Nations (2025). Ageing. *Global Issues*. Available at: <www.un.org/en/global-issues/ageing>.

Wacquant (2004/2009). *Punishing the Poor: The Neoliberal Government of Social Insecurity*. Trans. anonymous. Durham, NC: Duke University Press.

Wallace-Wells, D. (2019). *The Uninhabitable Earth: A Story of the Future*. London: Penguin Books.

Yorke, J. (2013). *Into the Woods: How Stories Work and Why We Tell Them*. London: Penguin Books.

Žižek, S. (2008). *Violence: Six Sideways Reflections*. New York: Picador Books.

Part III

A Solution

7 Counter

I opened this essay with my intention to demonstrate that cinematic fictions have the capacity to confront the climate challenge by shaping the desires of their audiences and then dividing this core claim into three constituent claims. The first, which was the subject of Part I, is that the climate challenge is a cultural rather than natural challenge (or, more accurately, *primarily* a cultural challenge). The third, which was the subject of Part II, is that Alfonso Cuarón's *Children of Men* (2006) is exemplary in confronting the climate challenge as a cultural challenge. The second, which is the subject of Part III and a synthesis of the first two parts, explains precisely how *Children of Men* in particular and cinematic fictions more generally contribute to confronting the cultural challenge. In Chapter 2, I suggested three approaches to cinematic fictions, beginning with green cultural criminology as a theoretical framework, moving on to the monomyth as a tool for both the creation and interpretation of Hollywood feature films, and concluding with a Marxist theory of aesthetic education. I provided a critique of these approaches in Chapter 3, identifying either a convincing objection to or a potential flaw in each. In this chapter, which introduces the final part of the essay, I address each of these objections and flaws, picking up where I left off in Chapter 3.

7.1 Representing Reality

Chapter 2 summarised the insights of green cultural criminology as follows (Brisman & South 2014): the way in which environmental harms are represented is crucial to their relation to the criminal justice system; the commodification of nature is one of the causes of harmful patterns of consumption; and the contestation of space is a successful means of resisting and ultimately reducing environmental harm. All three of representation, commodification, and contestation are sites for intervention in the politics of climate change, situating green cultural criminology in a broader approach in which the role of the social sciences is to counter contemporary attempts to depoliticise the climate crisis (Twine 2010, 2024). My summary also mentioned the recent focus on popular culture and fiction, about which green cultural criminology makes

DOI: 10.4324/9781003565826-10

two perspicacious claims (Brisman & South 2025): the dominant mode of the representation of nature is as one or more of monstrous, abject, or apocalyptic and this representation is harmful in virtue of the way in which it shapes the experiences, attitudes, and beliefs of audiences. As a theoretical framework, green cultural criminology acknowledges that fictions (and other popular cultural artefacts) can both misrepresent and represent and the evidence cited suggests that the former is the norm. Critiques of what I have previously called criminologies of narrative fiction, a category that overlaps with green cultural criminology, ask whether and how fictions can represent rather than simply reflect reality (Frauley 2010, 2021a, 2021b; McGregor 2021a). The argument is that fictions are works of make-believe, imagination, or invention and cannot tell us anything interesting about real people, places, and events. I think this view is naïve, demonstrating a lack of understanding about both representation and narrative.

A representation is something that stands for something else, for example: a portrait standing for a person living or dead, a flag standing for a nation state, a word on a page standing for an object in the world. If I type 'this page,' then 'this page' represents the page on which you are reading these words now, even though it has a number and appearance different to the page on which I typed 'this page' when I wrote this essay. My view, which is uncontroversial, is that a narrative is a representation. A narrative consists of two basic and essential components: *a sequence of events* and the *representation of those events* (McGregor 2021a, 2021b). This applies to the types of narratives that we refer to as fiction as well as those we refer to as nonfiction. This view is also uncontroversial. For example, an autobiography is not a life – it is a linguistic, verbal, or descriptive representation of a life by the liver of that life. Similarly, no representation of events includes every event that happened to a particular person or that occurred in a particular place. No matter how detailed and accurate an autobiography, it is impossible to represent every event in one's life. To take an everyday example, if later today my wife asks me what I did this morning, I wouldn't simply list every single action – and wouldn't know what to list if I did: every page typed, every word; every breath I took, every interval between breaths; every thought I had? I shall, instead, summarise all these events to coordinate them around one or two highlights or themes. My plan for this morning was actually to finish the first draft of Chapter 6, but towards the end of the second section, I started thinking about Chapter 7, typed the first draft of the paragraph you are reading now, and then returned to Chapter 6. My representation would include events relevant to this theme, such as interruptions or distractions or my decision to switch chapters. Narratives are thus both *selective*, representing some events but not others, and *unified*, selection being based on one or more themes (McGregor 2021a, 2021b). This also applies to both fiction and nonfiction narratives and is also uncontroversial.

My view is the moment one leaves an event out of a representation or selects a theme around which to organise that representation, there is a significant sense in which one fictionalises it. When I decide to relate the (nonfiction) narrative of my morning's work in terms of writing and thinking about Chapters 6 and 7, I represent the events in a way that communicates little more than my experience of those events. What actually happened to me this morning was simply a series of events – typing, thinking, drinking, breathing – one damn thing after another (typing then drinking) and some damn things simultaneously (typing and thinking). By selecting some of those thoughts and not others, like the thoughts about Chapters 6 and 7 rather than those about how many emails are waiting in my inbox and whether I will have time to go for a run this afternoon, I represent this small part of my life as being about writing this essay. Even if my answer to my wife's question is scrupulously honest, describing exactly how I experienced living through these hours, there is still potential to misrepresent the sequence of events. Maybe, for example, I actually spent more minutes thinking about emails and exercise than I did thinking about *Children of Men* and will only realise how poor my concentration was when I edit this chapter in a few days or weeks. My point is that the moment I impose narrative form on a sequence of events, there is a sense in which I fictionalise that sequence of events – because of selectivity and unity. This is a more controversial view, but also fairly obvious if one sets aside preconceptions about fiction and nonfiction and follows the logic of storytelling from creation to communication. To summarise what I have said so far, nonfiction narratives are more similar to fiction than we usually think.

Let us take a look from the other direction. A fictional narrative also has two components and they are the same components we have already discussed: a sequence of events and a representation of that sequence of events. One might ask how a made-up, imagined, or invented narrative has a sequence of events when we are considering characters, settings, and actions rather than people, places, and events. The sequence of events in *Children of Men* is everything that happens to every character from the moment the narrative opens (let us say Theo entering the coffee shop) to the moment the narrative closes (Kee seeing the Human Project boat). As the audience, we do not see everything that happens to every character in the course of the week (sequence of events) in the 100 minutes of the film (representation of the sequence of events). We do not even see everything that happens to Kee (as the protagonist) in that time or Theo (as the character with the most screen time). The sequence of events is thus imagined and the default setting for our imaginations is known as the reality principle: unless the cinematic narrative provides us with contrary information, we assume that Kee went to the toilet several times each day; ate several meals each day; slept for some hours each day; and had conversations with Miriam, Julian, Luke, Theo, Jasper, and others that were not part of the representation (Belsey 2011). Even though 2027

London is represented as being different to 2006 (and 2025) London, I assume the River Thames still runs through it, that it is still located in the south-east of England, and that it is still the biggest urban area in the United Kingdom. None of these features of the metropolis are stated in the narrative, but audiences import what they know about the world into the work – and, indeed, directors rely on audiences importing what they know because storytelling would be difficult if not impossible if they did not.

Children of Men is, as I noted in Part II, science fiction without advanced technology, but the importation to which I am referring applies to even the most fantastic fiction. I assume, for example, that the dragons in *Game of Thrones* fly in some vaguely similar way to birds, by which I mean that if one or more of their wings are injured, they will be unable to continue to defy gravity. I also assume that, like most animals, they cannot survive having their heads severed. Neither of these characteristics are made explicit in the narrative, but they are actually quite important because they contribute to my understanding that, as powerful as they are, adult dragons are neither invulnerable nor immortal. To summarise this less controversial side of my argument, fictional narratives are more similar to nonfiction than we usually think. Bringing these two sides together, my solution to the problem of representation is that where narratives are concerned, the difference between fiction and nonfiction is much less significant than assumed in consequence of the combination of selectivity and unity on the one hand and the reality principle on the other hand. In other words, representation in narrative fiction is no more complex or problematic than representation in narrative nonfiction, and one can apply whatever one's preferred theory of representation in narrative nonfiction is to narrative fiction.[1]

7.2 Monomythic Subversion

In Chapter 3, I summarised Joseph Campbell's (2008) concerns about the legacy of the monomyth in the 20th century following modernity's shift of meaning from the group to the individual and the subsequent appropriation of the monomyth by Hollywood (and other contemporary film and television industries). While the monomyth has always been the hero's journey, it was made meaningful by the hero's place in a group that was unified by one or more of a shared mythology, religion, or culture. In late capitalist civilisation, all we have left is capitalist realism, an ideology that promotes not only the pursuit of individual rather than collective projects, but also the competition between individuals as driving physical and psychological success (Harvey 2005; Hall 2011; Fisher 2022). I described the Hollywood blockbuster as *monomyth-lite*, a self-conscious product of the late capitalist imperative to maximise profit that seeks universal meanings and values which will appeal to all audiences but whose lack of specificity risks creating a narrative drained of significance. This presents two problems for the case I am trying to make for *Children of*

Men as an instrument of aesthetic education. First, it seems unlikely that the monomyth-lite, as a stripped-back and pared-down version of the monomyth, will have anything of value to say about a phenomenon as complex and urgent as the climate crisis. Second, even if the monomyth-lite could have something to say about phenomena as complex and urgent as the climate crisis, there seems to be an incompatibility between the monomyth itself and the challenge of climate change.

In his magisterial *Everything Must Go: The Stories We Tell About the End of the World*, Dorian Lynskey (2024: 373–374) sets the scope of the problem out both clearly and concisely:

> Compared to nuclear war, the climate emergency deprives popular storytellers of their usual toolkit. Global warming may move too fast for the planet but it is too slow for catastrophic fiction. How does one craft a tight plot out of a crisis that unfolds over decades rather than months or days? How can the heroes solve the problem and leave audiences with a cheering sense of closure? And who are the villains to be thwarted? Fossil-fuel tycoons and denialist politicians can wear the black hat but our collective complicity in a carbon-fuelled society is a buzzkill, morally and dramatically. What's more, the climate emergency is a result not only of the actions we have traditionally considered pernicious but of the things we valorize, like procreation, progress and the pursuit of a better standard of living.

Climate change is a global problem and action by a single – or even a coalition – of countries, religions, or cultures is not going to reduce or prevent ecocide. Ecocide requires global action and it is not clear how either the monomyth-lite or even the monomyth itself could contribute to a process of aesthetic education about the phenomenon.

My view is that the monomyth is indeed unsuited to meeting the challenge of climate change and I have watched with interest as novelists in particular experiment with different ways of writing about climate change. These experiments include, but are not limited to, subgenres such as planetary realism, the massively multi-protagonist novel, and novels by aliens and involve both innovation and the revival of earlier approaches to storytelling (Ganguly 2020; Googasian 2022; Marshall 2023). If the monomyth is an unsuitable form for dealing with climate change, then *Children of Men's* monomythic structure should undermine rather than facilitate its capacity for aesthetic education. I could respond by either denying that Cuarón uses monomythic storytelling or making the case that the film's monomythic structure is superficial, contributing little if anything to its more significant structure as a genuine allegory. Both would be disingenuous. In Chapter 4, I demonstrated the monomythic structure by using John Yorke's (2013) five point template, noting only that its architecture is more subtle that usual in a Hollywood feature film. Similarly, the significance of the structure to the allegory cannot be restricted to the literal level of meaning because the literal level underpins the other three representational levels of meaning and all four levels underpin the film's extra-representation meaning. Without the monomyth there is neither representation nor replication of living through the crisis. My solution to the problem is, instead, that Cuarón matches the monomyth to the climate crisis by subverting it.

I described this subversion in Chapter 6, in my discussion of two versions of the monomyth instantiated in *Children of Men*, a superficial one that constitutes what Slavoj Žižek (cited in *Children of Men* 2007) might call the foreground in his anamorphic reading of the film and a more substantive one that constitutes what he does call the background. *Children of Men* is apparently Theo's journey and, indeed, he appears in not only every scene but also every sequence onscreen. When characters leave his line of sight, they disappear from us as the audience and we do not know what happens to them or whether they will live or die (Miriam, Syd, and Marichka to take just three examples). Theo even remains onscreen after his death, when the Human Project boat arrives to pick up Kee and Dylan. My claim in Chapter 6 was that the superimposition of two monomyths at the literal level of meaning was an example of anamorphosis, as the foreground is used to bring the background into focus. The superimposition of Theo's journey over Kee's journey is both consistent with Žižek's anamorphic reading and the means by which Cuarón subverts the monomyth. In effect, he exploits it as what Gayatri Chakravorty Spivak (2012) refers to as a double bind (introduced in Chapter 3): monomythic storytelling is aimed at a global rather than local audience, but the globalisation has deprived it of what matters most (its mythic, religious, or cultural values) and the goal is to use it to reach a broad audience (and satisfy the demands of investors) while avoiding transforming it into monomyth-lite. Cuarón's strategy is to use Theo's heroic journey to achieve the former and Kee's to achieve the latter, attracting and engaging audiences with a familiar tale (increased by the fact that the hero is a White man) and then complicating that tale by revealing the second and more significant journey in the background. That the film was not a commercial success suggests that Cuarón erred on the side of avoiding monomyth-lite, telling a story that was too complex for his studio to sell.

7.3 Formal Complexity

In Chapter 2, I summarised Fredric Jameson's (1981, 1991, 2002, 2005, 2007, 2013, 2019) version of aesthetic education as the idea, hypothesis, or theory that films can provide utopian ways of thinking about the world in their form and style. I noted that it was composed of three interlocking concepts: capitalist realism, genuine allegory, and utopian desire. Genuine allegory reveals the structure of fourfold meaning within a text and in so doing creates meaning in excess of the representational capacity of the text and stages the complexity of social reality by holding a mirror and a microscope to everyday life. I explained these different levels and types of meaning as they are instantiated in *Children of Men* in detail in Chapter 6. Capitalist realism provides both the content and context of the film, which is a self-reflective product of and site of resistance to capitalist realism as collective ideology. Utopian desire is embodied by and in the cinematic narrative and the desire embodied by and in *Children of Men* is the desire for being-towards-ecocide, *re-attunement of the self and complete overhaul of the individual human experience with respect to the annihilation of our species*, as a new form of life or structure of feeling. Genuine allegories are thick narratives in which extra-representational meaning (the replication rather than representation of the apocalypse in *Children of Men*) cannot be explained by reducing it to its constitutive levels. Jameson (2019: 263) describes this narrative event as a spatial anomaly that creates 'a reading experience utterly unique and quite distinct from either conventional narrative or pure lyric.' The problem arises when the spatiality of the narrative event is contrasted with the temporality of the monomyth. If *Children of Men* has a narrative structure that is both monomythic and allegorical, then the two structures are not only in tension, but conflict: one draws the audience forward to the next shot, sequence, or scene and the other breaks that rhythm by revealing the architecture of apocalyptic replication. Any insight into ecocide, despair, capitalist realism, and/or the phenomenology of the apocalypse *Children of Men* has the potential to provide as an allegory will, it seems, be undermined at its very foundation by the functioning of the monomyth at the literal level of meaning.

The most obvious answer to this problem originates with Jameson himself, in a short monograph on Raymond Chandler's hardboiled detective fiction, in which he (Jameson 2016) evaluates what he refers to as Chandler's canon, the first four Philip Marlowe novels: *The Big Sleep*; *Farewell, My Lovely*; *The High Window*; and *The Lady in the Lake*. While criticising their content as clichéd and nostalgic – lacking in all political, social, or philosophical significance – Jameson argues that their form is surprisingly sophisticated and complex.[2] He (Jameson 2016: 22) identifies two distinct types of form in the novels, 'an objective form and a subjective one, the rigid external structure of the detective story on the one hand, and a more personal distinctive rhythm of events on the other'. The objective/subjective dichotomy is then revealed as supervenient on a temporal/spatial dichotomy. *Temporal form* consists of the structure of the detective story, the murder mystery plot, interrelationships of intrigue and action, and the logic of deduction employed by the detective. *Spatial form* consists of the arrangement of events, the descriptions of searches, the micro-episodic dimensions of the stories, and the relationships among episode-characters and episode-types. Jameson (2016: 24) describes Chandler's sleight of hand, in which each of the novels 'passes itself off as a murder mystery' when they are 'first and foremost descriptions of searches'.

Jameson maintains that they do not undermine each other because the spatial form is a product of the internal contradictions of the temporal form. The spatial form is in fact Chandler's inventive solution to a formal problem in detective fiction – the problem of how to exercise the reader's faculties of analysis and reasoning without facilitating the premature solution of the mystery. Jameson (2016: 86) refers to this solution as 'the ultimate secret of Chandlerian narrative.' In Chandler's canon, the murder victims are not merely already dead (killed prior to Marlowe's involvement), but *always already dead* (killed in the pasts of one or more of the main characters and barely remembered). In *The Big Sleep*, for example, Marlowe is hired to deal with the blackmail of Carmen Sternwood by her father and the actual murder mystery – the apparent disappearance of Rusty Regan – does not take centre stage until chapter 20 (of 32).

The temporal and spatial forms are thus complementary, two organising centres of the narrative in which each provides a lens through which the other is viewed (Jameson 2016: 27):

> The search and the murder serve as alternating centers for our attention in a kind of intricate Gestalt pattern; each serves to mask off the weaker, less convincing aspects of the other, each serves to arrest the blurring of the other out into the magical and the symbolic and to refocus it in a raw and sordid clarity.

When the reader's attention is focused on the murder, the search becomes a depressingly fatalistic circular narrowing down of options; when the reader's attention is focused on the search, the murder becomes a purposeless event, the senseless destruction of a lifeworld. Jameson's evaluation of Chandler's canon is in fact based on Martin Heidegger's (1950, 1927) elaborate and abstruse philosophy of art, expounded in his essay 'The Ontology of the Work of Art', which is based on the fundamental ontology he developed in *Being and Time*. One does not, however, need to employ or accept either Heidegger's fundamental ontology or philosophy of art to see how works and texts can have complex forms that complement rather than conflict with one another and exist in a state of creative rather than destructive tension. Jameson (2016: 76) summarises Chandler's 'formal peculiarities' as a literary design that (1) foregrounds dualisms while both (2) problematising and (3) reinventing them. Jameson's dichotomy of temporal and spatial form can be applied to *Children of Men* directly, to the monomythic and allegorical structures of the film respectively. When the audience's attention is focused on Theo's journey (or, perhaps on a second viewing, on Kee's journey), the crisis and its consequences become an all-too-familiar-capitalist reality; when the audience's attention is focused on the lived experience of the apocalypse, Theo or Kee's journeys become a call to action to reduce or prevent ecocide. Such an interpretation would also add yet another level of complexity to Žižek's anamorphic analysis of the film. Finally – and, I think – significantly, there is a sense in which all genuine allegories function in a way similar to that described above. Not all genuine allegories deploy a monomythic structure, but all of those with which Jameson (2019) is concerned are narratives and narrative is an essentially rather than accidentally temporal form. What is interesting and different about allegorical form is that it interrupts the flow from beginning to middle to end. The coexistence of spatial and temporal form in *Children of Men* is thus one of the features that enable its potential for aesthetic education.

Notes

1 My own preference is the ancient and until relatively recently widely acknowledged theory that fictions represent universals rather than particulars (see McGregor 2016, 2026).
2 While I agree with Jameson about the complexity of the form of Chandler's canon, I disagree about the simplicity of its content (see McGregor 2022).

References

Belsey, C. (2011). *A Future for Criticism*. Hoboken, NJ: Wiley-Blackwell.
Brisman, A. & South, N. (2014). *Green Cultural Criminology: Constructions of Environmental Harm, Consumerism, and Resistance to Ecocide*. Abingdon: Routledge.
Brisman, A. & South, N. (2025). *Monstrous Nature and Representations of Environmental Harms: A Green Cultural Criminological Perspective*. Philadelphia, PA: Temple University Press.
Campbell, J. (2008). *The Hero with a Thousand Faces*. 3rd ed. Novato, CA: New World Library.
Children of Men (2006). Directed by Alfonso Cuarón. US: Universal Pictures.
Children of Men – Comments by Slavoj Žižek (2007). In: *Children of Men – Special Edition*. Directed by Alfonso Cuarón. DVD, B000NIMZZM. Universal City, CA: Universal.
Fisher, M. (2022). *Capitalist Realism: Is There No Alternative?* 2nd ed. Alresford: Zero Books.
Frauley, J. (2010). *Criminology, Deviance, and the Silver Screen: The Fictional Reality and the Criminological Imagination*. New York: Palgrave Macmillan.
Frauley, J. (2021a). The Poverty of the Comparative Orthodoxy: Cultural Criminology, Perspectival Realism and Conceptual Variation. *International Journal of Law, Crime and Justice*, 66 (September), 1–12.
Frauley, J. (2021b). Fictional Realities and Criminology: Apprehending Social Reality Through Narrative Fiction. *Journal of Theoretical and Philosophical Criminology Special Edition: A Criminology of Narrative Fiction*, 13 (November), 111–125.

Ganguly, D. (2020). Catastrophic Form and Planetary Realism. *New Literary History*, 51 (2), 419–453.
Googasian, V. (2022). Feeling Fictional: Climate Crisis and the Massively Multi-Protagonist Novel. *New Literary History*, 53 (2), 197–216.
Hall, S. (2011). The Neo-Liberal Revolution. *Cultural Studies*, 25 (6), 705–728.
Harvey, D. (2005). *A Brief History of Neoliberalism*. Oxford: Oxford University Press.
Heidegger, M. (1927/1962). *Being and Time*. Trans. J. Macquarrie & E. Robinson. New York: Harper Collins Publishers.
Heidegger, M. (1950/2011). The Origin of the Work of Art. Trans. anonymous. In: *Basic Writings: Martin Heidegger*. Abingdon: Routledge, 89–139.
Jameson, F. (1981/2002). *The Political Unconscious: Narrative as a Socially Symbolic Act*. New York: Routledge.
Jameson, F. (1991). *Postmodernism, or, the Cultural Logic of Late Capitalism*. Durham, NC: Duke University Press.
Jameson, F. (2002). *A Singular Modernity: Essay on the Ontology of the Present*. London: Verso Books.
Jameson, F. (2005). *Archaeologies of the Future: The Desire Called Utopia and Other Science Fictions*. London: Verso Books.
Jameson, F. (2007). *The Modernist Papers*. London: Verso Books.
Jameson, F. (2013). *The Antimonies of Realism*. London: Verso Books.
Jameson, F. (2016). *Raymond Chandler: The Detections of Totality*. London: Verso Books.
Jameson, F. (2019). *Allegory and Ideology*. London: Verso Books.
Lynskey, D. (2024). *Everything Must Go: The Stories We Tell about the End of the World*. London: Pan Macmillan.
Marshall, K. (2023). *Novels by Aliens: Weird Tales and the Twenty-First Century*. Chicago: Chicago UP.
McGregor, R. (2016). *The Value of Literature*. London: Rowman & Littlefield International.
McGregor, R. (2021a). *A Criminology of Narrative Fiction*. Bristol: Bristol University Press.
McGregor, R. (2021b). *Critical Criminology and Literary Criticism*. Bristol: Bristol University Press.
McGregor, R. (2022). The Complex Art of Murder. *Journal of Aesthetic Education*, 56 (3), 63–80.
McGregor, R. (2026). *Reducing Political Violence: Narrative, Critique, and Criminology*. Bristol: Bristol University Press.
Spivak, G.C. (2012). *An Aesthetic Education in the Era of Globalization*. Cambridge, MA: Harvard University Press.
Twine, R. (2010). *Animals as Biotechnology: Ethics, Sustainability and Critical Animal Studies*. Abingdon: Routledge.
Twine, R. (2024). *The Climate Crisis and Other Animals*. Sydney: Sydney University Press.
Yorke, J. (2013). *Into the Woods: How Stories Work and Why We Tell Them*. London: Penguin Books.

8 Configuration

This chapter concludes both the third part of the essay and the essay itself and is where I draw on the work of previous chapters to demonstrate that Alfonso Cuarón's *Children of Men* (2006) shapes the desires of its audiences by means of its configuration of the cinematic experience. The shaping of desires is one of the ways in which the film has the capacity to confront the climate crisis as a cultural crisis and this potential to influence audiences is recognised by and in green cultural criminology as a theoretical framework and the broader tradition of aesthetic education within which I have located that framework in this essay.

If *Children of Men* can shape the desires of its audiences then so can other films and popular cultural artefacts. I have focused exclusively on *Children of Men* because the challenge it presents to the climate crisis is exemplary and because this is a short monograph that could not accommodate analyses of other exemplars, such as Octavia Butler's (1993, 1998) *Parables* and HBO's *Game of Thrones* (2011–2019). The shaping of desire (as well as attitudes and beliefs) by means of experiential configuration is precisely what Avi Brisman and Nigel South (2025) recognise when they argue that the representation of nature as monstrous, abject, or apocalyptic is harmgenic. Narrative fictions and other kinds of popular cultural artefacts can both increase and reduce harm in consequence of their impact on audiences. One cannot have it one way without the other: if fiction represents reality, which it does, then it can also misrepresent reality. My focus here has been on representation and this essay is exclusively about the potential to reduce harm and thus in contrast with the majority of criminological research on fiction to date.

8.1 Political Power

I want to begin this ending with a broader and potentially much more substantial objection to my claims about *Children of Men*, drawn from an unusual source, a long review of a short book. Mark Bould's (2021) controversial but compelling *The Anthropocene Unconscious: Climate Catastrophe Culture* is a monograph written in response to Amitav Ghosh's (2016) *The Great*

DOI: 10.4324/9781003565826-11

Derangement: Climate Change and the Unthinkable. Ghosh's monograph is based on the Randy L. and Melvin R. Berlin Lectures he delivered at the University of Chicago in 2015 and divided into three parts: Stories, History, and Politics.[1] Although the monograph is well worth reading in its entirety, there are two aspects of the first part that are relevant to my argument in this essay. The first of these concerns the failure of the institution of literature to respond to the challenge of climate change and I shall discuss the second in the final section of this chapter. Ghosh's (2016: 79; see also Bould 2021) *Great Derangement* is a play on the 'Great Acceleration', the dramatic and synchronous increase in a range of socioeconomic trends from population and gross domestic product to transportation and telecommunications since 1945. The derangement is contemporary culture's lack of ability or will to address the climate change caused by the Great Acceleration. Ghosh imagines a future in which critics identify the paradox of a culture that both prided itself on its self-awareness and ignored its self-destruction. Ghosh's thesis begins with the claim that literature has never recovered from the split into literary and genre fiction, which he dates to the publication of Mary Shelley's *Frankenstein; or, The Modern Prometheus*, in 1818. He then argues that while genre – science, speculative, or climate – fiction has attempted to meet the challenge of climate change, it has for the most part failed in virtue of being nothing more than disaster fiction set in the future. Ghosh (2016: 76) concludes that literary or artistic fiction has not even tried to meet the challenge and cites Abdul Rahman Munif's 1984 'Petrofiction' novel, حلملا ندم (*Cities of Salt*), as one of the exceptions that proves the rule. The Hollywood film industry has been accused of a similar derangement with, for example, Rewrite the Future director Daniel Hinerfeld (cited in Bauck 2023) claiming that 'Hollywood has not reflected the greatest drama of our generation'.

Bould (2021) begins *The Anthropocene Unconscious* with the provocative claim that *all* cultural texts are about the Anthropocene, albeit usually on an unconscious level. This is based on a for the most part implicit and unacknowledged appropriation of Fredric Jameson's (1981) conception of the political unconscious, which serves as a precursor to his *Poetics of Social Forms* and a prototype of his delineation of genuine allegory. Jameson makes the case that all texts are political because, first, they are all produced and consumed within the historical context of class struggle between oppressor and oppressed and, second, that this struggle permeates both the form and the content of the stories human beings tell themselves and one another. Bould makes a similar move and given that the popularisation of the novel as a literary form was contemporaneous with the Industrial Revolution, there is a sense in which first novels and later feature films and television series were all produced and consumed within the Anthropocene. He claims, in consequence, that in the same way that the class struggle permeates all texts to characterise them as political so the Anthropocene permeates all texts to characterise them as ecological.[2] Bould (2021: 103) evinces his thesis by delineating the 'environmental uncanny', already articulated by Ghosh (2016: 32), as involving both the recognition by human beings that they are in the presence of nonhuman agency and an anthropocentric hubris which assumes it will be able to control its nonhuman creations. As such, Bould turns Ghosh's concept against him: the environmental uncanny is everywhere, from Jane Austen's *Mansfield Park* to Richard Powers' *The Overstory* and from Mauro Herce's *Dead Slow Ahead* to the *Fast & Furious* and *Sharknado* film franchises.[3]

In an incisive and comprehensive review essay of *The Anthropocene Unconscious* in the *Ancillary Review of Books*, Fabius Mayland (2022) divides culture into two orders. First-order culture is fiction about climate change, whether the extent to which it is actually about climate change is disputed (such as Austen's *Mansfield Park*) or not (such as Munif's *Cities of Salt*). Second-order culture is '*culture that is already in the business of observing culture*' or theory about fiction about climate change (Mayland 2022, emphasis in original. Mayland (2022) then accuses Bould of naivety, of imagining that 'climate politics just needs a little more imagination and storytelling to really get going.' Uncontrovertibly, what humanity needs is to avert, avoid, or ameliorate the climate crisis. Mayland maintains that neither more fiction about climate change (first-order culture) nor more theory about fiction about climate change (second-order culture, where Bould fits in) can achieve this aim. Both the crisis and its solutions have been established beyond all reasonable doubts by the natural sciences and what is required is the political power to implement them. In Chapter 1, I used the phrase *tragedy of climate change science*, Bruce Glavovic, Timothy Smith, and Iain White's (2022) description of the failure of governments to act on the evidence of climate change at the same time as they acted on the evidence of the COVID-19 pandemic. Like Glavovic, Smith, and White, Dorian Lynskey (2024: 369, emphasis in original) draws attention to the length of time that governments and their populations have known about the phenomenon:

> The phrase *climate crisis* first appeared in 1981. *Climate emergency* followed in 1989, when a record 79 percent of Americans said that they had some awareness of the greenhouse effect. In 1990, the IPCC published its first report and new generation of activists mobilized 200 million people worldwide for the biggest Earth Day to date. In 1992, world leaders signed the significant but non-binding United Nations Framework Convention on Climate Change in Rio de Janeiro.

Lynskey (2024: 370) notes first that the potential breakthrough in mobilising a meaningful response to climate change never happened and then that since 'the Rio Declaration in 1992, we have pumped more carbon into the atmosphere than in the preceding thirty thousand years of human history.' Lynskey, Glavovic, Smith, White, and Mayland all provide evidence for my initial claim that the climate challenge is (primarily) a cultural challenge, a problem for social rather than natural scientific investigation.

Mayland's (2022, emphasis in original) answer to this problem is:

> not academic theory but naked political power. Is our culture today 'about' climate change? Bould's account is, I think, more convincing than that of Ghosh. Nor is the question necessarily an uninteresting one. We in the humanities should be honest enough, however, to admit that the *stakes* of answering this question are low either way.

What he means by stakes is that neither second-order culture (academic theory) nor first-order culture (literature and film) makes much difference either way and Mayland (2022) elaborates on Bould's naivety by referring to his 'overblown belief in the power of literature', which is essentially rather than accidentally shared by all humanities' scholars, who must justify the salaries and grants they receive. Mayland's criticism certainly resonates. I have wondered many times about the point of writing this essay and why I am doing it. I don't need it for my career so is it just an excuse to spend time with a favourite film or favourite theorists? I can probably never be sure and have reflected on the role of cultural criticism and cultural critics elsewhere (McGregor 2024). Notwithstanding, Mayland misses something very important about his first-order culture. Stories of or about climate change *do* matter. Avi Brisman (2012, 2023), who has also commented on Ghosh and Bould, notes both the significance of stories to climate change denial and the apparent lack of storytelling expertise of natural scientists. The deniers, he suggests, tell better stories and his own take on climate change as a cultural challenge is climate change as a battle of competing narratives. Lynskey (2024: 373) reiterates the point when he states: 'It might seem trivial to worry about the dearth of compelling stories about climate change, but the failure of fiction to get a handle on the subject reflects the struggle of scientists to craft an unignorable narrative.' I shall suggest precisely how fictions matter in the final section of this chapter, but if first-order culture matters, then so does second-order culture – even if for no other reason than to demonstrate why and how first-order culture matters.

8.2 Postcapitalist Culture

While I agree with Mayland that naked political power is important, it is not the most important component of meeting the climate challenge as a cultural challenge. My rationale is that it would be naïve to imagine that the cultural changes required to address the challenge could be achieved exclusively from the top down. To return to Glavovic, Smith, and White's (2022) comparison of climate change with the COVID-19 pandemic, the pandemic demonstrated the capacity of proportions of the global population to resist even the most benign official impositions – vaccinations, mask-wearing, and lockdowns – where a meaningful response to the climate crisis would require the radical alteration of lifestyles on a permanent basis, that is the transformation of our forms of life or structures of feeling, as suggested in Chapter 1 (Wittgenstein 1953; Williams 1958). I return to this point shortly, but even if the global population could be coerced into new forms of life or structures of feeling by national governments without mass violence, it is highly unlikely that those governments would take voluntary action for the simple reason that they are already too invested in what Jason Moore (2015) refers to as capitalist world-ecology to begin dismantling it. This is of course the value of Mark Fisher's (2022b) capitalist realism, which draws attention to the fact that neither populations nor governments can imagine an alternative to capitalist world-ecology. The failure of the imagination is both top down and bottom up and raises the question of whether meeting the challenge of climate change requires a transformation of world-ecology.

Moore thinks so. So do Kehinde Andrews (2021) and Ariel Salleh (2017), in spite of approaching climate change from different perspectives, antiracism and feminism respectively. Richard Twine (2024) reaches the same conclusion from a critical animal studies perspective and Naomi Klein (2014) has demonstrated the lengths to which global capital has gone to depoliticise the climate crisis as a crisis of capitalist world-ecology. There is a growing consensus among social scientists that the challenge of climate change cannot be met without transforming contemporary world-ecology and the pertinent question is precisely what needs to be dismantled – corporate capitalism (Whyte 2020), neoliberal capitalism (Harvey 2005), or capitalism itself (Žižek 1999). The central thesis of *Literary Theory and Criminology* is that all three of the mass harms of ecocide, racism, and sexism are linked by global capitalism and that the causal relation offers an unprecedented opportunity to reduce all three (as well as other harms) at once rather than in a piecemeal fashion (McGregor 2023a). I cited Klein (2014: 10) in support of this idea:

> I am convinced that climate change represents an historic opportunity on an even greater scale. As part of the project of getting our emissions down to the levels many scientists recommend, we once again have the chance to advance policies that dramatically improve lives, close the gap between rich and poor, create huge numbers of good jobs, and reinvigorate democracy from the ground up.

Not only can the challenge of climate change not be met exclusively from the top down – because there will be violent resistance to naked political power, even when the exercise of that power is in a just cause – but it is unlikely to be met from the top down in consequence of those in a position to exercise such power being for the most part already part of the problem, i.e. the socioeconomic elite of global capitalism. Mobilisation from the bottom up is thus at least as important and likely more important and climate cultures need to be changed in populations with the power to take action, vote, and resist unsustainability. The question is whether that is even possible in our world-ecology, a culture of capitalist realism.

Although Fisher's (2022b) long essay is written in an optimistic (albeit at times outraged) tone, its purpose is to diagnose, examine, and critique capitalist realism as an all-pervasive and almost invisible collective ideology. As such, he does not have much to say about a remedy or resistance, about how populations and governments can be persuaded that there are alternatives to capitalist world-ecology and that a just and sustainable world-ecology is not merely a utopian desire incapable of realisation. Critique is, of course, much easier than praxis, pointing out what is wrong with existing systems, institutions, and practices relatively simple when compared to recommending replacements that will actually be more just and sustainable (Harcourt 2020). Fisher's (2018) blueprint for resistance, the working title of which was *Acid Communism*, was in progress at the time of his death and the unfinished introduction was published in *k-punk: The Collected and Unpublished Writings of Mark Fisher (2004–2016)*. I do not find it particularly enlightening and there is no reason it should be as both a relatively small part of the proposed book and incomplete. While also incomplete, a more promising place to start is *Postcapitalist Desire: The Final Lectures*, the bulk of which is a transcript of the 5 (of a planned 15) lectures Fisher (2022a) delivered before his death. In a posthumous tribute to Fisher (Fisher et al. 2017: 13, emphasis in original), his colleagues introduce the course (and thus the subsequently published book) as follows:

> Built upon the intricately sketched landscape of Capitalist Realism, at the heart of the naturalised order of appearances assumed to render all alternatives impossible, 'Post-Capitalist Desire' is a climax in Mark's commitment to envision a future for the left. It calls into question capital's long-established monopoly on desire.

In the first lecture, 'What is Postcapitalism?', Fisher (2022a: 39) sets out the success of capitalist world-ecology in terms of its monopolisation and commodification of desire, commenting on a televised critique of the Occupy movement in 2010:

> There's a narrative behind it, which is a story about desire. These protesters have the products of advanced capitalism, therefore . . . it's not only that they're hypocrites, it's that they don't really want what they say they want. They don't really want a wealth beyond capitalism. What they want is all of the fruits of capitalism – and ultimately that's why capitalism will win. They may claim, ethically, that they want to live in a different world

> but libidinally, at the level of desire, they are committed to living within the current capitalist world.

By drawing attention to social media and smartphones as products of capitalist world-ecology, Fisher raises the question of degrowth: would postcapitalism require some kind of slowing or shrinking of the economy and is that something people will ever desire? In his second lecture, he (Fisher 2022a: 83) introduces Sigmund Freud as 'a thinker of the historicity of desire' because if desire is historical rather than natural, it is contingent rather than necessary and thus capable of being shaped or transformed in the future. In the third lecture, Fisher (2022a: 126) draws on both Immanuel Kant and György Lukács to reiterate the relation between capitalist world-ecology and desire:

> Capital is purposiveness without purpose. Endless driving . . . there's no final purpose to it. There's no end point to it, in itself, which I think brings us very close to the core theme of this module, in a way. Because you could say that makes it flat with the structure of desire in itself.

Capitalism and desire are so intervolved that their structures are indiscernible or inseparable and Fisher continues by comparing the mysteries of the Roman Catholic Church in the Middle Ages to the mysteries of the late capitalist economy in the 21st century. Neither were intended to be either understood or questioned; both required, instead, absolute faith and an imagination with clearly defined limits.

After discussing capitalist world-ecology's success in appropriating and commodifying resistance, counterculture, and revolution in the fourth lecture, Fisher (2022a: 176–177) draws on Jean-François Lyotard, points to the apparent absence of desire in Marxism, and explains why it is an impediment to postcapitalism:

> If desire is monopolised by capital, as we suggested at the start, then . . . that's it then, isn't it? The changes we need are never going to happen. Or rather, we have this vision of a kind of political project, on the one hand, and desire on the other, so that, in order to achieve this political project, we'll have to subdue our own desires. And that is highly problematic, I would suggest.

In the final lecture, 'Libidinal Marxism', Fisher (2022a: 194–195, emphasis in original) sketches the form postcapitalist desire might take without specifying its content:

> There is no beyond that is untainted by capital, which is different from saying that things always have to be as they are now. The implicit message, surely, is that we have to imagine a transformation *out of* where we

> are now. We can't fall for any temptation to look for an untainted region, a non-alienated region, etc. We have to start from a full immersion in capital.

Ultimately, *Postcapitalist Desire* is, like *Acid Communism*, a disappointment, because both were works in progress that were interrupted while specifying the problem and before the development of a solution. My purpose in summarising Fisher here is first to elaborate on my previous discussions of capitalist realism and second to suggest the significance of the concept of postcapitalist desire to meeting climate change as a cultural challenge.

8.3 Shaping Desire

In Part II, I interpreted and appreciated *Children of Men* in terms of Jameson's (1981, 1991, 2002, 2005, 2007, 2013, 2019) *Poetics of Social Forms*, as providing new ways of thinking about the world in its form and style. As such, the film embodies the utopian desire for being-towards-ecocide, re-attunement and overhaul of individual human experience with respect to the annihilation of our species (Burns 2023). As a genuine allegory, *Children of Men* embodies this desire as part of a complex structure, consisting of multiple levels of representational and extra-representational meaning and realising multiple values. As discussed in Chapter 6, the literal level of meaning is movement from an inaugural condition in which a cynical, dispassionate Theo leads a meaningless life to his decision to dedicate himself to Kee's protection when he discovers her pregnancy to a retrospectively inevitable condition in which Theo makes his life meaningful by sacrificing it for the future of humanity. The form and style of *Children of Men* invite the audience to approve of this movement and to sympathise, empathise, or identify with Theo's progress. The film also complicates the hero's journey by revealing, at precisely the point of retrospective inevitability, that there is a simultaneous and deeper meaning – a movement from an inaugural condition in which an immature and isolated Kee is a victim of circumstance to her decision to trust Theo to save her and her child to a retrospectively inevitable condition in which her courage and competence facilitate the redemption of humanity. Again, the film invites the audience to sympathise, empathise, or identify with Kee's progress. If, in other words, we are sensitive and responsive to Cuarón's storytelling, we will desire Theo's recreation of meaning and Kee's maturation and be satisfied by a resolution in which both are achieved, albeit perhaps with some regret that Theo fails to see Kee's rescue.

The literal level of meaning underpins the symbolic, existential, and anthropic meanings of *Children of Men*, which present us with a complex and compelling set of relations among Theo and Kee's respective journeys, the fictional fertility crisis as representing the reality of ecocide, the conflict between hope and despair in response to these crises, and the unfolding of both the crises and responses to them within the limitations of capitalist realism. In each case, Cuarón uses form and style to configure our attitudes towards these meanings, encouraging us to reject ecocide, despair, and capitalist realism in both the world of the film and the world in which we watch it. All four of these levels of representational meaning in turn underpin the extra-representational meaning, an allegorical interruption of narrative temporality in which the spatial relations between the characters and the audience are revealed as a parallel experience of living in and through a crisis of and in the present. Again, audience experience is carefully configured in terms of attitudes and desires as we recognise, perhaps with horror, that there is little significant difference between the lives of the characters onscreen and our own lives and that both are in an equally significant sense dystopian. Like all great works of art, *Children of Men* creates an experience that engages our emotions and imaginations to shape our beliefs, attitudes, and desires. There is of course no guarantee that we will be affected by the film in the manner I have described because we may either be too insensitive or too inattentive to understand the various levels of meaning or because we may understand them but resist them for ethical, political, or religious reasons. If one is, however, receptive to the experience Cuarón has curated, then we also experience an at least temporary configuration of our beliefs, attitudes, and – most importantly – desires.

Returning to the debate between Ghosh (2016) and Bould (2021), my inclination is to disagree with both of them: I don't see how one could seriously argue that fiction has ignored ecocide without establishing a rigid divide between literary and popular fiction, which Ghosh rightly wants to avoid, and if Bould wants to convince us that ecocide is ubiquitous in popular and literary fiction, then he needs to make more of a case for its similarity to class conflict. Perhaps such criticism misses precisely the point. Ghosh and Bould are both making deliberately provocative, controversial, and polemic claims in order to impel their readers to think carefully about an extremely important question, the relation between art and life in what I have called the Capitalocene and the role of art in education about the Capitalocene. Ghosh's monograph is longer and narrower in scope than Bould's and discusses this relation and role in more detail. Ghosh is at his most insightful when delineating the problem, which in his case means explaining the Great Derangement. By way of introduction, he (Ghosh 2016: 9, emphasis in original) identifies the climate crisis as a crisis of both culture and imagination: 'Indeed, this is perhaps the most important question ever to confront *culture* in the broadest sense – for let us make no mistake: the climate crisis is also a crisis of culture, and thus of the imagination.' I explained the conception of the climate crisis as a crisis of culture in Part I and as a crisis of imagination – in terms of the imagination required to represent it onscreen and within the confines of monomythic storytelling – in Part II.

The paragraph that follows is worth quoting in full (Ghosh 2016: 9–10, emphasis in original):

> Culture generates desires – for vehicles and appliances, for certain kinds of gardens and dwellings – that are among the principal drivers of the carbon economy. A speedy convertible excites us neither because of any love for metal and chrome, nor because of an abstract understanding of its engineering. It excites us because it evokes an image of a road arrowing through a pristine landscape; we think of freedom and the wind in our hair; we envision James Dean and Peter Fonda racing toward the horizon; we think also of Jack Kerouac and Vladimir Nabokov. When we see an advertisement that links a picture of a tropical island to the word *paradise*, the longings that are kindled in us have a chain of transmission that stretches back to Daniel Defoe and Jean-Jacques Rousseau: the flight that will transport us to the island is merely an ember in that fire. When we see a green lawn that has been watered with desalinated water, in Abu Dhabi or Southern California or some other environment where people had once been content to spend their water thriftily in nurturing a single vine or shrub, we are looking at an expression of a yearning that may have been midwifed by the novels of Jane Austen. The artifacts and commodities that are conjured up by these desires are, in a sense, at once expressions and concealments of the cultural matrix that brought them into being.

Culture generates patterns of desire by its impact on our beliefs, attitudes, emotions, and imaginations and works of art achieve the same end on a complex level, generating desires internal to the work (e.g. my desire that Theo and Kee achieve their shared goal) and external to the work (e.g. my desire to confront the climate crisis and to do so in hope rather than in despair). In consequence of this complexity, the impact of the desires generated by art on human behaviour is much more difficult to measure than the impact of desires generated by the advertising Ghosh mentions. My opinion is that it is impossible to measure these desires in a meaningful way, but those of us who engage with and are responsive to works of art have first-hand experience of their existence (McGregor 2026). Culture, art, popular culture, literature, and films not only generate new desires but also shape existing desires.

My co-author David Grčki concluded *An Epistemology of Criminological Cinema* by noting the significance of desire (as well as emotion and passion) to understanding the cinematic experience and recommending the *oeuvre* of Jacques Lacan as a way of understanding desire. He (Grčki & McGregor 2024: 132) makes the controversial claim that 'Lacan's theory encompasses human irrationality, our needs, demands, desires, and fantasies effectively, and has a theory of cinematic engagement.' Whether this is an accurate evaluation of Lacan or not, the relation between art and desire is both important and opaque. We know little more than Ghosh states, that desire is shaped by culture and that literature and film configure desire. In the previous section, I claimed that Mayland (2022) exhibited the same naivety of which he accuses Bould (2021) in suggesting that political power is all that is required to confront climate change as a cultural challenge. My reasons were twofold: that the people with the capacity to exert influence from the top down were unlikely to do so and that even if they did, it would be resisted from the bottom up. The former may well be an unfair criticism of Mayland because he is identifying top down power as the solution to the problem and is not required to explain how this might be brought about (certainly not in a review essay). Ultimately, however, bottom up is at least as important as top down, if for no other reason than sheer weight of numbers. Extinction Rebellion staged its most recent (and possibly last) mass protest in the United Kingdom in April 2023. The aim was to bring 100,000 people to demonstrate at the Houses of Parliament in London for four days. The organisation claimed that 60,000 took part, but even if the goal had been achieved, 100,000 is not even two-tenths of a single per cent of the UK's population (Hudson 2023; McGregor 2023b). Imagine if a million people had turned up. Imagine seven million, over 10 per cent of the population. Change would happen and it would happen quickly.

Ghosh (2016) and Brisman (2023) both identify the climate crisis as a crisis of culture and a crisis of imagination. It is also a crisis of desire. If we cannot imagine an alternative to capitalist realism, then we cannot desire it and if we cannot desire it, we have no desire on which to act. Films like *Children of Men* help us not only imagine life beyond capitalist realism but also generate a desire for it and it is the latter which is, I think, the more important of the two. The human population is full of geniuses who can imagine things the rest of us cannot, but if we do not desire what they imagine, then that genius is unlikely to be realised. I close with a poignant and poetic quote from the penultimate paragraph of Bould (2021: 141–142):

> Imagine waking from this anthropogenic nightmare.
> Imagine possibility.
> Imagine a world of trees, of all the other lives with which they make kin, of rich and healthy biodiversity.
> Imagine a world where energy is not just clean, but free. Where no one lacks water, food, shelter, community, health.
> Imagine plenitude without profligacy, produced here without creating poverty there. Imagine this Red Plenty, and turn it Green.
> Imagine replacing separation, isolation, hostility and suspicion with a Red Belonging that says without hesitation or cavil 'it doesn't matter where you come from or who you are, we will care for you anyway.'
> And imagine extending that unconditional care – that respect and love – to the biosphere . . .

Notes

1 To further complicate provenance, a part of Ghosh (2016) was subsequently published as Ghosh (2021) in Penguin's *Green Ideas* series.
2 'Ecological' is my paraphrase; Bould (2021: 3) actually uses being '"about" climate change.'
3 As controversial as this may (or may not) sound, Timothy Morton (2018, 2021) has a very similar and highly popular thesis about art (including literature and film). In a manner similar to Bould, Morton's thesis also relies on an implicit and unacknowledged affinity with Gayatri Chakravorty Spivak's (1999, 2012, 2014) contribution to the tradition of aesthetic education.

References

Andrews, K. (2021). *The New Age of Empire: How Racism and Colonialism Still Rule the World*. London: Allen Lane.

Bauck, W. (2023). 'The Visuals of Today Help Create the Reality of Tomorrow:' Why Hollywood is Finally Tackling Climate Change Onscreen. 28 February. *Fast Company*. Available at: <www.fastcompany.com/90856208/the-visuals-of-today-help-create-the-reality-of-tomorrow-why-hollywood-is-finally-tackling-climate-change-onscreen>.

Bould, M. (2021). *The Anthropocene Unconscious: Climate Catastrophe Culture*. London: Verso Books.

Brisman, A. (2012). The Cultural Silence of Climate Change Contrarianism. In: White, R. (ed.). *Climate Change from a Criminological Perspective*. New York: Springer, 41–70.

Brisman, A. (2023). Ecocide and *Khattam-Shud*. *Journal of Aesthetic Education Special Issue: Aesthetic Education through Narrative Art*, 57 (3), 107–123.

Brisman, A. & South, N. (2025). *Monstrous Nature and Representations of Environmental Harms: A Green Cultural Criminological Perspective*. Philadelphia, PA: Temple University Press.

Burns, R. (2023). From Meaning to Ecocide: The Value of Phenomenology for Green Criminology. *Critical Criminology: An International Journal*, 31 (4), 1137–1154.

Butler, O.E. (1993/2019). *Parable of the Sower*. New York: Grand Central Publishing.

Butler, O.E. (1998/2019). *Parable of the Talents*. London: Headline Publishing Group.

Children of Men (2006). Directed by Alfonso Cuarón. US: Universal Pictures.

Children of Men – Comments by Slavoj Žižek (2007). In: *Children of Men – Special Edition*. Directed by Alfonso Cuarón. DVD, B000NIMZZM. Universal City, CA: Universal.

Fisher, M. (2018). *k-punk: The Collected and Unpublished Writings of Mark Fisher (2004–2016)*. London: Repeater Books.

Fisher, M. (2022a). *Postcapitalist Desire: The Final Lectures*. London: Repeater Books.

Fisher, M. (2022b). *Capitalist Realism: Is There No Alternative?* 2nd ed. Alresford: Zero Books.

Fisher, M., Barcelos, L., Colquhoun, M., Eastwood, M., Eshun, K., Moalemi, M. & Ronkina, G. (2017). The Fisher-Function. *Place of Publication Unknown: Egress*. Available at: <https://s3.amazonaws.com/arena-attachments/1713222/4a82190d0905bf8e000df1899d22a993.pdf>. Accessed 31.10.24.

Game of Thrones (2011–2019). Originally released 17 April. US: HBO.

Ghosh, A. (2016/2017). *The Great Derangement: Climate Change and the Unthinkable*. Chicago: University of Chicago Press.

Ghosh, A. (2021). *Uncanny and Improbably Events*. London: Penguin Books.

Glavovic, B.C., Smith, T.F. & White, I. (2022). The Tragedy of Climate Change Science. *Climate and Development*, 14 (9), 829–833.

Grčki, D. & McGregor, R. (2024). *An Epistemology of Criminological Cinema*. Abingdon: Routledge.

Harcourt, B.E. (2020). *Critique and Praxis: A Critical Philosophy of Illusions, Values, and Action*. New York: Columbia University Press.

Harvey, D. (2005). *A Brief History of Neoliberalism*. Oxford: Oxford University Press.

Hudson, M. (2023). Extinction Rebellion Gave It 'the Big One' with a Four-Day Peaceful Protest – Now What? 27 April. *The Conversation*. Available at: <https://theconversation.com/extinction-rebellion-gave-it-the-big-one-with-a-four-day-peaceful-protest-now-what-204581>.

Jameson, F. (1981/2002). *The Political Unconscious: Narrative as a Socially Symbolic Act*. New York: Routledge.

Jameson, F. (1991). *Postmodernism, or, the Cultural Logic of Late Capitalism*. Durham, NC: Duke University Press.

Jameson, F. (2002). *A Singular Modernity: Essay on the Ontology of the Present*. London: Verso Books.

Jameson, F. (2005). *Archaeologies of the Future: The Desire Called Utopia and Other Science Fictions*. London: Verso Books.

Jameson, F. (2007). *The Modernist Papers*. London: Verso Books.

Jameson, F. (2013). *The Antimonies of Realism*. London: Verso Books.

Jameson, F. (2019). *Allegory and Ideology*. London: Verso Books.

Klein, N. (2014). *This Changes Everything: Capitalism vs. the Climate*. London: Penguin Books.

Lynskey, D. (2024). *Everything Must Go: The Stories We Tell about the End of the World*. London: Pan Macmillan.

Mayland, F. (2022). Climate Change Lurking Behind Every Corner: Review of Mark Bould's. In: *The Anthropocene Unconscious: Ancillary Review of Books*. 11 March. Available at: <https://ancillaryreviewofbooks.org/2022/03/11/climate-change-lurking-behind-every-corner/>.

McGregor, R. (2023a). *Literary Theory and Criminology*. Abingdon: Routledge.

McGregor, R. (2023b). The World-Ecology of Climate Change Cinema. *Theaker's Quarterly Fiction*, 75, 63–71.

McGregor, R. (2024). Making Cultural Criticism Matter. *International Journal for the Semiotics of Law*. DOI: 10.1007/s11196-024-10192-6.

McGregor, R. (2026). *Reducing Political Violence: Narrative, Critique, and Criminology*. Bristol: Bristol University Press.

Moore, J. (2015). *Capitalism in the Web of Life: Ecology and the Accumulation of Capital*. New York: Verso Books.

Morton, T. (2018). *Being Ecological*. London: Pelican Books.

Morton, T. (2021). *All Art is Ecological*. London: Penguin Books.

Salleh, A. (2017). *Ecofeminism as Politics: Nature, Marx and the Postmodern*. 2nd ed. London: Zed Books.

Spivak, G.C. (1999). *A Critique of Postcolonial Reason: Toward a History of the Vanishing Present*. Cambridge, MA: Harvard University Press.

Spivak, G.C. (2012). *An Aesthetic Education in the Era of Globalization*. Cambridge, MA: Harvard University Press.

Spivak, G.C. (2014). *Readings*. London: Seagull Books.

Twine, R. (2024). *The Climate Crisis and Other Animals*. Sydney: Sydney University Press.

Whyte, D. (2020). *Ecocide: Kill the Corporation before It Kills Us*. Manchester: Manchester University Press.

Williams, R. (1958). *Culture and Society*. New York: Columbia University Press.

Wittgenstein, L. (1953/2009). *Philosophical Investigations*. Trans. G.E.M. Anscombe, P.M.S. Hacker & J. Schulte. Chichester: Wiley-Blackwell.

Žižek, S. (1999/2009). *The Ticklish Subject: The Absent Centre of Political Ideology*. London: Verso Books.

Index

For Product Safety Concerns and Information please contact our EU representative GPSR@taylorandfrancis.com
Taylor & Francis Verlag GmbH, Kaufingerstraße 24, 80331 München, Germany

www.ingramcontent.com/pod-product-compliance
Lightning Source LLC
LaVergne TN
LVHW010934110826
845149LV00013B/2597

* 9 7 8 1 0 3 2 9 3 4 3 0 3 *